Protocols used in Molecular Biology

Edited by

Sandeep Kumar Singh

Indian Scientific Education and Technology Foundation, Lucknow-226002, India

Centre of Biomedical Research, SGPGI Campus, Lucknow-226014, India

&

Dhiraj Kumar

Rameshwar College, B.R.A. Bihar University, Muzaffarpur-842001, India

Protocols Used in Molecular Biology

Editor: Sandeep Kumar Singh

ISBN (Online): 9789811439315

ISBN (Print): 9789811439292

need for a court order if at any point you breach any terms of this License Agreement. In no event will any delay or failure by Bentham Science Publishers in enforcing your compliance with this License Agreement constitute a waiver of any of its rights.

3. You acknowledge that you have read this License Agreement, and agree to be bound by its terms and conditions. To the extent that any other terms and conditions presented on any website of Bentham Science Publishers conflict with, or are inconsistent with, the terms and conditions set out in this License Agreement, you acknowledge that the terms and conditions set out in this License Agreement shall prevail.

Bentham Science Publishers Pte. Ltd.
80 Robinson Road #02-00
Singapore 068898
Singapore
Email: subscriptions@benthamscience.net

CONTENTS

Divakar Singh, Tarun Minocha, Satyavrat Tripathi, Rupika Sinha, Shubhankar Anand, Hareram Birla, Vivek Kumar Pandey, Arun Rawat, Smita Gupta, Sanjeev Kumar Yadav, Pawan Kumar Dubey and *Pradeep Srivastava*

FOREWORD

Inside the knowledge that includes every living cell, the molecular biology is the science that takes as aims the study of the processes that are developed in the living beings from a molecular point of view. In a modern sense, the molecular biology tries to explain the phenomena of life in terms of their macromolecular properties. Two macromolecules, in particular, are its study object: the nucleic acids, the most used being Deoxyribonucleic acid (DNA), the main component of genes, and the proteins, which are the active agents of the living organisms. Inside the Project Human Genome, that was a project of scientific research with the fundamental target to determine the sequence of pairs of chemical bases that compose DNA and to identify approximately 20.000-25.000 genes of the human genome from a physical and functional point of view; it is possible to define the molecular biology as the study of the structure, function and composition of the molecules that are biologically important.

Hence, this area is related to other fields from biology to chemistry, and particularly genetic and biochemical engineering. The molecular biology concerns principally the understanding of the interactions of different systems in the cell, wherein it includes many relations, among them, between DNA with RNA, the synthesis of proteins, the metabolism, and how all these interactions are regulated to afford a correct functioning of the cell.

Therefore, the present book entitled *"Protocols used in Molecular Biology"* mainly includes the most recent and advance molecular biology protocols with a concise introduction, materials and chemicals required, step-by-step procedure, the trouble shooting tips and finally the applications of protocols. This protocol book includes essential techniques of proteomics, genomics, cell culture, epigenetic modification and structural biology with a special focus on the fundamental applications of each protocol, which are lacking in most of the protocol books. This protocol book will be significantly important for the academicians, molecular biologists, graduates and undergraduate students engaged in basic and clinical research. Protocols of this book can be utilized to unravel the problem of cancer biology, genetics, neuroscience and many more. In particular, a neuroscientist can utilize this protocol book to isolate RNA and DNA from brain tissues.

In this sense, we believe that the knowledge that will be acquired by the reading of the chapters contained in this book edited by Prof. Sandeep Kumar Singh, and the later discussions on every topic, will give to the readers an important and valuable tool to understand the different biological and molecular methods that take place in the organism and the functionalities that make a clear differentiation between every organ and each basic molecular structure of our human body.

Prof. Dr. Eduardo Sobarzo-Sánchez
Laboratory of Pharmaceutical Chemistry, Faculty of Pharmacy,
University of Santiago de Compostela,
Spain

&

Prof. Dr. Seyed Mohammad Nabavi
Applied Biotechnology Research Center,
Baqiyatallah University of Medical Sciences,
Tehran
Iran

PREFACE

Understanding biological phenomena, diagnosis of diseases, elucidating the key target molecule underlying a particular disease and developing new therapeutic approaches strongly rely on the foundation of advanced protocols frequently used in molecular biology experiments. Keeping this background the present book entitled "Protocols used in Molecular Biology" mainly includes the most recent and advance molecular biology protocols with a concise introduction, materials and chemicals required, step-by-step procedure, the trouble shooting tips and applications of protocols. The book includes essential techniques of proteomics, genomics, cell culture, epigenetic modification and structural biology with a special focus on the fundamental applications of each protocol. This protocol book will be significantly important for the academicians, molecular biologists, graduates and undergraduate students at basic and clinical research. Protocols of this book can be utilized to unravel the problem of cancer biology, genetics, neuroscience and many more. In particular, a neuroscientist can utilize this protocol book to isolate RNA and DNA from brain tissues. The expression of genes at mRNA level using real time polymerase chain reaction and at protein level using a modified western blot protocol for enhanced sensitivity in the detection of tissue protein can be studied. Furthermore, neuroscientists can also study the region specific expression of mRNA following RNA in situ hybridization and protein by immunohistochemistry. Several environmental factors can cause changes in DNA without changing its sequence through epigenetic modifications, which can be studied through sodium bisulfite conversion of genomic DNA. Such environmental factors can also alter the neuronal architecture, which can be easily elucidated by rapid Golgi method to unravel the change in the number, length and arborisation of dendritic spine. Furthermore, cell culture and live cell imaging protocols can be particularly utilized to understand the changes at cellular level in various pathological conditions. Differential change in the proteome of neurodegenerative diseases can be studied through 2D DIGE. A change in point mutation of the brain can be elucidated through Mis-match Amplification Mutation Assay (MAMA). The DNA and protein complex formation essentially regulate either an increase or a decrease in the expression of gene during pathological conditions. Studying such complexes through electrophoretic mobility shift assay (EMSA) can open a new avenue for understanding complexes and regulation of gene expression. Thus, this protocol book will be a complete package for molecular biologists working in various fields.

Sandeep Kumar Singh
Indian Scientific Education and Technology Foundation,
Lucknow-226002,
India

&

Dhiraj Kumar
Kala Azar Medical Research Centre,
Rambag Road, Muzaffarpur-842001,
India

List of Contributors

Arshad Md Molecular Endocrinology Lab, Department of Zoology, University of Lucknow, Lucknow, India

Anand Shubhankar School of Biochemical Engineering, Indian Institute of Technology (Banaras Hindu University), Varanasi-221005, India

Arun Rawat Department of Biochemistry, Institute of Science, Banaras Hindu University, Varanasi-221005, India

Birla Hareram Department of Biochemistry, Banaras Hindu University, Varanasi-221005, India

Bharati Shikha School of Life Science, Jawaharlal Nehru University, New Delhi, 110067, India

Bhatt Madan Lal Brahma Department of Radiotherapy, King George's Medical University, Lucknow-226003, India

Dinesh Raj Modi Department of Biotechnology, Babasaheb Bhimrao Ambedkar University (A Central University), Vidya Vihar, Raibareilly Road, Lucknow-226025 (U.P.), India

Gupta Smita Department of Microbiology, Institute of Medical Science Banaras Hindu University, Varanasi-221005, India

Gupta Shalini Department of Oral Pathology and Microbiology, King George's Medical University, Lucknow-226003, India

Gedda Mallikarjuna Rao Department of Biochemistry, Institute of Science Banaras Hindu University, Varanasi-221005, India

Jafri Asif Molecular Endocrinology Lab, Department of Zoology, University of Lucknow, Lucknow, India

Jai Godheja School of Life and Allied Sciences, ITM University, Atal Nagar, Raipur, Chhattisgarh-492001, India

Kumar Sudhir Molecular Endocrinology Lab, Department of Zoology, University of Lucknow, Lucknow, India

Kanakala Surapathrudu Institute of Plant Sciences, Agricultural Research Organization, Volcani Center, Israel

Kavyanjali Sharma Department of Pathology, Faculty of Medicine, Banaras Hindu University, Varanasi. U.P, India

Kumar Bapatla Kesava Pavan Institute of Plant Sciences, Agricultural Research Organization, Volcani Center, Israel

Misra Anamika Institute of medical Science, Banaras Hindu University, Varanasi-221005, India

Mishra Aastha CSIR-Institute of Genomics and Integrative Biology, Delhi, India

Modi Arpan Institute of Plant Sciences, Agricultural Research Organization, Volcani Center, Israel

Naik Aijaz A. School of Studies in Neuroscience, Jiwaji University, Gwalior-474011, India

Pasha Qadar	CSIR-Institute of Genomics and Integrative Biology, Delhi, India
Pathak Abhishek	Department of Neurology, Institute of Medical Sciences Banaras Hindu University, Varanasi, India
Pawan Kumar Dubey	Centre for Genetic Disorder, Institute of Science, Banaras Hindu University, Varanasi-221005, India
Pradeep Srivastava	School of Biochemical Engineering, Indian Institute of Technology, Banaras Hindu University, Varanasi-221005, India
Rais Juhi	Molecular Endocrinology Lab, Department of Zoology, University of Lucknow, Lucknow, India
Rai Sachchida Nand	Department of Biochemistry, Banaras Hindu University, Varanasi-221005, India
Singh Abhishek	School of Life Sciences, Jawaharlal Nehru University, New Delhi, India
Singh Divakar	School of Biochemical Engineering, Indian Institute of Technology (Banaras Hindu University), Varanasi-221005, India
Shakuntala Mahilkar	Hepatitis Division, National Institute of virology, Pune, Maharashtra, India
Singh Pallavi	Dr. A.P.J. Abdul Kalam Technical University, Lucknow, Uttar Pradesh, India
Srivastava Pradeep	School of Biochemical Engineering, Indian Institute of Technology (Banaras Hindu University), Varanasi-221005, India
Singh Payal	Department of Biochemistry, Institute of Science Banaras Hindu University, Varanasi-221005, India
Singh Surya Pratap	Department of Biochemistry, Institute of Science Banaras Hindu University, Varanasi-221005, India
Pandey Surya Prakash	Department of Neurology, Institute of Medical Sciences Banaras Hindu University, Varanasi, India
Sinha Rupika	School of Biochemical Engineering, Indian Institute of Technology (Banaras Hindu University), Varanasi-221005, India
Singh Rakesh K.	Department of Biochemistry, Institute of Science Banaras Hindu University, Varanasi-221005, India
Shekhar Sudhir K	Department of Biotechnology, Babasaheb Bhimrao Ambedkar University (A Central University) Vidya Vihar, Raibareilly Road, Lucknow-226025 (U.P.), India
Singh Saumitra Sen	Department of Biochemistry, Institute of Science Banaras Hindu University, Varanasi-221005, India
Sanyal Somali	Amity Institute of Biotechnology, Amity University, Uttar Pradesh, Lucknow 226028, India
Sanjeev Kumar Yadav	Department of Zoology, Institute of Science, Banaras Hindu University, Varanasi-221005, India
Smita Gupta	Department of Biochemistry, Institute of Science, Banaras Hindu University, Varanasi-221005, India
Subash Chandra Sonkar	Department of Obstetrics and Gynecology, Vardhman Mahaveer Medical College and Safdarjung Hospital, Ansari Nagar New Delhi- 1100029, India

Tripathi Madhu   Molecular Endocrinology Lab, Department of Zoology, University of Lucknow, Lucknow, India

Tiwari Neeraj   Department of Biochemistry, Institute of Science Banaras Hindu University, Varanasi-221005, India

Tripathi Satyavrat   School of Biochemical Engineering, Indian Institute of Technology Banaras Hindu University, Varanasi-221005, India

Tarun Minocha   Department of Zoology, Institute of Science, Banaras Hindu University, Varanasi-221005, India

Vivek Kumar Pandey   Centre for Genetic Disorder, Institute of Science, Banaras Hindu University, Varanasi-221005, India

Yadav Suresh Kumar   Amity Institute of Biotechnology, Amity University, Uttar Pradesh, Lucknow 226028, India

Zahra Walia   Department of Biochemistry, Institute of Science Banaras Hindu University, Varanasi-221005, India

CHAPTER 1

Isolation of Genomic DNA From Plant Tissues

Pallavi Singh[*]

Dr. A.P.J. Abdul Kalam Technical University, Lucknow, Uttar Pradesh, India

Abstract: Genomic DNA extraction is the starting point for various downstream molecular biology applications *viz.* PCR, restriction analysis, hybridisation *etc.* Numerous problems like DNA degradation, co-isolation of viscous polysaccharides, polyphenols and other secondary metabolites causing damage to DNA, inhibiting restriction enzymes, DNA polymerases *etc*, are routinely encountered during DNA isolation from plants. Quinone compounds resulting from oxidation of polyphenols lead brown the DNA preparations and can also damage proteins and DNA's due to their oxidizing properties. This results in a poor yield of high molecular weight DNA. The protocol below explains the extraction of DNA *via* the CTAB method, involving three major steps *viz* lysis of cell wall and membranes, extraction of genomic DNA and precipitation of DNA.

Keywords: DNA isolation, CTAB, Plants, Genomic DNA, Plant Tissues.

INTRODUCTION

The isolation of pure DNA is the first important step in the process of molecular studies in plants. The isolated DNA should be suitable for digestion using restriction endonucleases. Some plants are notorious due to their intractability. With many isolation techniques, one has to modify the protocols for each plant species depending on diversity and their biochemical composition particularly secondary metabolites. The method described by Doyle and Doyle [1] is used as the most successful protocol in many plant species.

Principle

Cetyl-Trimethyl-Ammonium–Bromide (CTAB) is a cationic detergent with chemical structure as given beside. This has many useful properties, which makes it one of the main chemicals that produce a large number of polysaccharides while purifying DNA from plants [2]. It tends to form complexes with proteins and acidic polysaccrides, in solutions of high ionic strength, but doesn't precipitates

---

[*] **Corresponding author Pallavi Singh:** Dr. A.P.J. Abdul Kalam Technical University, Lucknow Uttar Pradesh, India; E-mail: pallavibiotech@yahoo.com

Sandeep Kumar & Dhiraj Kumar (Eds.)

the nucleic acids present in it [3]. While in low ionic strength conditions it precipitates proteins along with some polysaccharides. Under these conditions, proteins and neutral polysaccharides remain in the solution. CTAB extraction buffer with high ionic strength (1.4M NaCl) is added to homogenised plant cell lysates. It forms a complex with polysaccharides/proteins, which can be sequentially extracted with chloroform and phenol and genomic DNA is recovered from the supernatant by precipitating with isopropanol or ethanol [4].

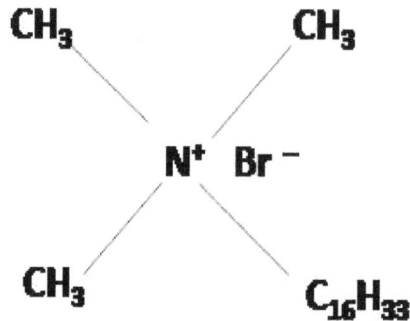

$$CH_3 \qquad CH_3$$
$$N^+ \quad Br^-$$
$$CH_3 \qquad C_{16}H_{33}$$

Sample Collection and Storage Conditions

Although, choice of fresh plant tissue is ideal for the genomic DNA isolation, but it's not always possible. Thus in that case frozen samples stored at -20°C for short periods or -70°C or lower temperature (liquid nitrogen -196°C) for longer periods can also be used. Thawing of the frozen samples should be avoided as the sub cellular disruption while freezing, leads to rapid degradation of DNA in thaw samples due to increased nuclease activity. Fresh samples kept for a couple of days in a refrigerator or cold room (4°C) or dry plant material (may be from herbarium samples) can also use in DNA extraction. The latter helps in collection and storage of a large number of samples at a very at low cost [5].

1. **Lysis of Cell Membrane**: The main step of extraction which involves the rupture of cell and nucleus membrane/wall is achieved with help of detergent (CTAB), used in the extraction buffer along with Tris-HCl and EDTA. In the presence of specific NaCl concentration CTAB captures the lipids and proteins which help in easier release of genomic DNA, forming a insoluble complex with CTAB. EDTA on the other hand is a chelating agent, which helps in reducing the DNAse activity after binding to its cofactor magnesium. Tris-HCl is a buffering agent which helps to maintain the pH of the extraction buffer as low or high pH can damage the DNA. When all the cell organelles are broken apart in the solution along with the genomic DNA, purification of the latter is performed.

2. **Extraction**: In this step, all the contaminants (polysaccharides, phenolic

compounds, proteins and cell debris) are separated from CTAB-nucleic acid complex formed above, with the help of Phenol and chloroform. Chloroform helps Phenol in the denaturation and removal of proteins from crude cell lysates. Also, it facilitates in clear formation of aqueous, polar phase at the top (containing the nucleic acid and water) and organic phase (containing the proteins and other cell component) at the bottom. Now, after the DNA is purified from other unwanted contaminants it can easily be precipitated.

3. **Precipitation**: In this final step, absolute alcohol helps in separation of DNA and CTAB complex, as the latter is more stable in alcohol rather than water. Thus, as the detergent is washed off, DNA gets precipitated. Further, the precipitated DNA is washed twice/thrice with 70% alcohol for further removal of any kind of salt attached with nucleic acid. The DNA pellet is air dried and dissolved in T10E1 buffer and kept -20°C /-80°C to for long term usage.

Materials Required Sample tissue, Liquid nitrogen, Sterile pestle and mortar, Sterile spatulas, scissors, tissue paper, Water bath (65°C), Sterile eppendorf tubes, Reagents for 3% CTAB extraction buffer, β-mercaptoethanol, Chloroform: Isoamyl alcohol solution(24:1), Phenol: Chloroform: Isoamyl alcohol solution (25:24:1), Chilled isopropanol, Chilled absolute ethanol, 70% ethanol, T10E1 buffer. Preparation of working solution is given in Table **1**.

Table 1. Preparation of working solutions.

Solutions	Method of Preparation
1M Tris	121.1 g Tris BASE (pH 8 adjusted with HCl) H_2O 800 ml make up the volume to 1000ml, Autoclave.
0.5M EDTA (pH8.0)	EDTA 186.1g (pH 8 adjusted with NaOH) H_2O 800ml make up the volume to 1000ml, autoclave.
4M NaCl	233.7 g of NaCl H_2O 800ml make up the volume to 1000ml, autoclave.
3M Sodium Acetate	Dissolve 24.60 g of Sodium Acetate in 60 ml of H_2O (pH to 5.2 with glacial acetic acid) make up the volume to 100 ml with H_2O. Autoclave.

CTAB DNA Extraction Buffer: 100 ml

100 mM Tris (pH 8.0) 10 ml of 1 M stock

20 mM EDTA 4 ml of 0.5 M stock

1.4 M NaCl 35ml of 4M stock

3% CTAB 3g

10 mM β-mercaptoethanol (Use 14.3 M stock) 70 μl

Make up the volume to 100 ml

$T_{10}E_1$ Buffer

10Mm Tris 500μl

1mM EDTA 100 μl

Make up the volume to 50ml with sterile distilled water

DNA Isolation Protocol

1. Take 2-3 small emerging leaves, and grind it to a fine powder using liquid nitrogen with the help of mortar and pestle.

2. Transfer the resulting powder into a 2 ml sterile centrifuge tube containing 1 ml DNA extraction buffer (100mM Tris HCl, pH 8; 20mM EDTA; 1.4 M NaCl ; 3% w/V CTAB; 0.2% β mercaptoethanol) and incubate the homogenate in water bath at 65°C for 30 min.

3. When incubation is over, take out the tubes and bring down to room temperature. Add 500μl chloroform: isoamyl alcohol (24:1), mix well with gentle inversions and centrifuge the tubes at 12,000 rpm for 15 min.

4. Take out supernatant in a fresh tube, add a double volume of chilled isopropanol, and mix gently with quick inversions. Allow the DNA to precipitate at -20°C for overnight / -80°C for 1 hour.

5. Spin the tube at 5,000 rpm for 15 min, discard the supernatant. Wash the DNA pellet thrice with 70% alcohol at the same rpm, air dry, suspend in 200 μl $T_{10} E_1$ buffer and treat with 3 μl RNase (10 mg/ml) at 37°C for 30 min.

Purification

6. Add 200 μl phenol: chloroform: isoamyl alcohol (25:24:1) mix and spin at 12,000 rpm for 10 min.

7. Take out supernatant in a fresh tube and add 200 μl of chloroform: isoamyl alcohol (24:1) spin at 12,000 rpm for 10 min.

8. Take out supernatant in fresh tube {add 1/10 volume 3M sodium acetate (pH 5.2) optional} and 500 μl of chilled absolute ethanol to precipitate DNA at -20°C for 1 hour.

9. Spin at 5,000 rpm for 10 min, discard the supernatant and air dry the DNA pellet.

10. Wash the DNA pellet thrice with 70% ethanol, dry and finally dissolve in 20-30 µl of $T_{10} E_1$ buffer and store at -20°C.

Useful Tips and Suggestions for Obtaining Optimum Results

- Transfer the ground material in liquid nitrogen immediately to the extraction buffer. Using excess liquid nitrogen for grinding is not necessary and is actually an inefficient way to grind. Use only the required volume to keep the tissue frozen. Fine grinding can be done only when there is a small amount of tissue in the frozen state.
- Trying to grind a large amount of sample is a bad idea. This will result in coarse grinding and will greatly reduce DNA yields.
- Care should be taken while removing the aqueous phase from the organic phase during chloroform: isoamyl alcohol extraction, to maximize the full recovery of DNA. If no separation is observed between the two phases, which may be due to the high concentration of DNA and/or cell debris in an aqueous phase, dilution with more buffer and re-extraction is recommended or suggested.
- Preferably, carry out all the operations in cold to keep a check over nucleases which, if released from a plant cell can lead to degradation of the genomic DNA and hinder the process.
- All the other operations should be as gentle as possible to ensure that the DNA is not shared. Do not vortex at any stage.
- It is advisable to preheat the extraction buffer to 65°C before starting the isolation process. This is because the CTAB-nucleic acid complex may precipitate prematurely during isolation. Therefore, except for the last two steps, do not refrigerate the samples.
- The DNA should not be overdried because it will then take longer time to re-suspend.

CONSENT FOR PUBLICATION

Not applicable.

CONFLICT OF INTEREST

The author confirms that this chapter contents have no conflict of interest.

ACKNOWLEDGEMENTS

Declared none.

REFERENCES

[1]   Doyle JJ, Doyle JL. A rapid DNA isolation procedure for small quantities of fresh leaf tissue. Phytochem Bull 1987; 19: 11-5.

[2]   Thomson D, Henry R. Use of DNA from dry leaves for PCR and RAPD analysis. Plant Mol Biol Report 1993; 11(3): 202-6.
[http://dx.doi.org/10.1007/BF02669845]

[3]   Jones AS, Walker RT. Isolation and analysis of the deoxyribonucleic acid of *Mycoplasma mycoides* var. *Capri.* Nature 1963; 198: 588-9.
[http://dx.doi.org/10.1038/198588a0] [PMID: 13964684]

[4]   Wilson K. Preparation of genomic DNA from bacteria. Curr Protoc Mol Biol 2001; 2(1): 4.
[PMID: 18265184]

[5]   Fallah F, Minaei Chenar H, Amiri H, *et al.* Comparison of two DNA extraction protocols from leave samples of Cotinus coggygria, Citrus sinensis and Genus juglans. Cell Mol Biol 2017; 63(2): 76-8.
[http://dx.doi.org/10.14715/cmb/2017.63.2.11] [PMID: 28364796]

CHAPTER 2

RNA Isolation Protocol from Cells and Tissues

Pallavi Singh[*]

Dr. A.P.J. Abdul Kalam Technical University, Lucknow, Uttar Pradesh, India

Abstract: The preparation of intact ribonucleic acid is difficult because of the action of nucleases, which are liberated upon tissue homogenisation. In many cells, high concentrations of the ribonucleases are reserved in the secretory granules and upon disruption of the cell, they get mixed with the RNA and lead to its degradation. Guanidinium chloride and thiocyanate are potent chaotropic agents that reduce hydrophobic interactions and disrupt protein tertiary structures, disassociate protein-nucleic acid complexes and disintegrate cellular structures. Guanidinium thiocyanate is especially strong protein denaturant because both the cation and anion disrupt the hydrophobic bonds between the amino acid side chains. RNA usually binds to proteins within a cell and this agent disassociates the nucleoprotein complex, without disrupting RNA structure. Thus RNA can be obtained by using these agents, after homogenisation and low-speed centrifugation and precipitated with ethanol. The protocol below explains the stepwise isolation of total RNA from cells and tissues using TRIzol reagent which is the mono-phasic solution of phenol and guanidine thiocyanate.

Keywords: Centrifugation, CTAB, DEPC, Formaldehyde Gel, Guanidinium chloride, Guanidine thiocyanate, Homogenization, Hydrophobic, Hot Phenol, Nucleoprotein, Nucleic acid, Protein, phenol, RNA isolation, TRIzol.

INTRODUCTION

Accuracy of transcriptome expression analysis depends on the purity of RNA samples. Due to high secondary metabolite contaminants, acidic nature of the cell sap and highly reactive nature of RNA, isolation of intact total RNA from some plants is sometimes challenging. The quality of isolated mRNA can indirectly be assayed with the following features; (i) Intact 28S, 18S rRNA and 5S rRNA bands (eukaryotic samples), (ii) The intensity of 28S rRNA band should be approximately twice as the 18S rRNA band, (iii) Partially degraded RNA will have a smeared appearance lacking the sharp rRNA bands. Completely degraded RNA will appear as a very low molecular weight smear. The total RNA can be extracted by several protocols described by many authors, but here basic Guanid-

[*] **Corresponding author Pallavi Singh:** Dr. A.P.J. Abdul Kalam Technical University, Lucknow Uttar Pradesh, India; E-mail: pallavibiotech@yahoo.com

ine and Non- Guanidine based RNA isolation protocols are described.

Precautions to be Taken Before the Experiment

RNAses can be introduced accidentally into the RNA preparation at any point in the isolation procedure by an improper technique. Because RNAse activity is difficult to inhibit, it is essential to prevent its introduction.

The following guidelines should be observed when working with RNA.

- Always wear disposable gloves. The skin often contains bacteria and moulds that can contaminate an RNA preparation and be a source of RNases.
- Use sterile, disposable plasticware and pipettes reserved for RNA work to prevent cross-contamination with RNases from shared equipment.
- In the presence of TRIzol reagent, RNA is protected from RNase contamination. Glass items can be baked at 150°C for 4 hours, and plastic items can be soaked for 10 min in 0.5 M NaOH, rinsed thoroughly with water, and autoclaved.
- A large amount of liquid nitrogen is needed during the experiment, sampling as well as extraction.
- Always try to immerse the tissue sample into liquid nitrogen soon after detachment from the plant, to minimize the expression of new mRNAs (which can occur within 5 min).
- Slice the bulky tissues samples into small pieces for storage in liquid nitrogen to avoid RNA degradation.

I. Isolation of RNA Using TRIzol Method

Principle

TRIzol combines phenol and guanidine thiocyanate, developed by Chomczynski and Sacchi [1], which facilitates the immediate and most effective inhibition of RNase activity. This reagent can simultaneously isolate RNA, DNA and protein from the plant, animal, human, yeast and bacterial sample. After the biological sample is homogenized and lysed in trizol containing guanidinium isothiocyanate (powerful protein denaturant) inactivating the RNase activity, acidic phenol/chloroform leads to partitioning into the aqueous phase and an organic phase. Due to its acidic nature, RNA partitions into the aqueous phase from which RNA can be precipitated by adding isopropanol and subsequently washed with 70% ethanol, and re-dissolve pellets into diethylpyrocarbonate (DEPC) water. The DNA and protein remains in the interphase and phenol phase respectively, which can be precipitated in successive steps.

Materials Required

TRIzol Reagent, Chloadecroform, Isopropanol, Phenol, 100% Ethanol, 75% Ethanol (in DEPC-treated water), Nuclease-free water, RNase-free DNase,3M sodium acetate, DEPC water, Mortar and pestle(pre-chilled, DEPC treated), 1.5ml Eppendorf tubes and tips (DEPC treated), liquid nitrogen.

Steps

1. Take 100 mg tissue, homogenate to powder with liquid nitrogen and transfer the powder to sterile Eppendorf tube.
2. Add 1ml of TRIzol reagent. Incubate at room temperature for 5 min.
3. Add 0.2 ml chloroform/1 ml TRIzol.
4. Shake tubes vigorously for 15 secs.
5. Incubate at room temperature for 3 min.
6. Centrifuge at 10,000 rpm for 15 min at $4°$ C.
7. RNA is in the top (aqueous phase, about 60% volume), transfer the RNA aqueous phase to a fresh tube.
8. Add 0.5 ml isopropanol/ml TRIzol to precipitate RNA.
9. Incubate for 10 min at room temperature.
10. Centrifuge at 10,000 rpm for 10 min at $4°$ C.
11. Remove and dispose of supernatant.
12. Wash RNA pellet with 75% ethanol.
13. Vortex and centrifuge at 8,000 rpm for 2 min at $4°$ C.
14. Air dry the RNA pellet.
15. Dissolve RNA in 100-200 μlRNase-free water.

II. Isolation of RNA Using the CTAB Method

This method is appropriate for all the problematic tissues which are rich in polysaccharides [2].

Materials Required

- Sterile tips and tubes (as required), RNase-free water, Chloroform: isoamyl alcohol (24:1).
- 12 M LiCl.
- **Extraction buffer**: 2% Cetyl tri-methyl-ammonium bromide (CTAB); 2% polyvinylpyrrolidone K30, 100 mM Tris-HCl, pH 8.0, 25 mM EDTA pH 8.0; 2 M NaCl, 0.5 g/L spermidine, and 2% β-mercaptoethanol. Autoclave, after dissolving in RNase free water.
- **Sodium dodecyl sulfate–Tris-HCl–EDTA (SSTE) buffer**: 1 M NaCl; 0.5% sodium dodecyl sulfate (SDS), 10 mM Tris-HCl, pH 8.0, and 1 m MEDTA,pH

8.0. Autoclave, after dissolving in RNase free water.

Steps

1. Grind the tissue in liquid nitrogen, thaw it completely and add to 15 ml of preheated (65°C) extraction buffer in an RNase free falcon tube (avoid the formation of lumps by vortexing).
2. Add 300 μl of β-mercaptoethanol. Vortex for few mins and keep at room temperature.
3. Add an equal volume of Chloroform: isoamyl alcohol (C:I) solution, vortex for mixing and transfer the components to 50 ml Oakridge tubes.
4. Balance all the samples and centrifuge at 10,000 rpm for 10 min at RT.
5. Carefully transfer the upper aqueous phase with the help of pipette, to a fresh tube without touching or disturbing the below interface material.
6. Repeat the C:I step again and take the upper aqueous layer in a fresh falcon tube.
7. Add 0.2 volumes of freshly prepared 12 M LiCl solution to it.
8. Incubate the solution at 4°C for overnight.
9. Centrifuge at 10,000 rpm for 10 min at 4°C.
10. Discard the supernatant completely, and add 200 μl preheated (65°C) SSTE buffer and transfer to1.5ml tube.
11. If some white precipitation is seen, keep the tube at 37°C until dissolved completely.
12. Add an equal volume of C: I solution. Vortex for mixing.
13. Add the double volume of absolute ethanol for precipitation of RNA and incubate (-80°C for 30 min or -20°C for 2 hours).
14. Centrifuge the tube at full speed for 15 min at 4°C.
15. Discard the supernatant completely, air dry the pellet and dissolve in 20 μl of nuclease-free water. Keep it on ice before preserving.

III. Hot-Phenol RNA Extraction Method

Materials Required

Extraction Buffer Make 50 ml Extraction buffer as follows

1M Tris-HCl (pH = 8.0): 5 ml

8M LiCl: 0.625 ml

0.5M EDTA (pH = 8.0): 1.0 ml

SDS: 0.5 gm

DEPC treated water: 43.2 ml

Steps

1. Add 1:1 volume of extraction buffer and phenol and heat at 80°C.
2. Grind approximately 100 mg of frozen tissue in Eppendorf tubes or in pestle and mortar with liquid nitrogen.
3. Add 500 µl of hot phenol + extraction buffer.
4. Vortex for 30 sec and add 250 µl C: I and re-vortex.
5. Centrifuge for 15 min at 10,000 rpm. Remove the aqueous phase and mix with 1/3 volume of 8M LiCl. Mix well by inverting the tubes.
6. Precipitate overnight at 4°C and centrifuge for 15 min at 10,000 rpm.
7. Re-dissolve the pellet in 250 µl DEPC water; add 0.1 volume of 3M sodium acetate (pH5.2).
8. Precipitate the RNA with a double volume of absolute ethanol.
9. Centrifuge for 15 min at 14,000 rpm.
10. Wash pellet with 70% ethanol and air dry the pellet.
11. Dissolve in 20 µl of nuclease-free water. Keep it on ice before preserving.

IV. QIAGEN RNeasy Kit Method

1. Wash young fresh leaf samples of the plant with 75% ethanol and freeze immediately in liquid nitrogen after harvesting.
2. Grind the required quantity of leaves into fine powder in liquid nitrogen with the help of pre-chilled pestle and mortar.
3. Transfer the powder to 2 ml collection tube containing 450 µl RLT buffer, vortex vigorously and incubate at room temperature for 5 min.
4. Centrifuge at 7000 rpm, for 5 min and transfer the upper aqueous layer to QIA shredder spin column placed in 2 ml collection tube.
5. Centrifuge for 2 min at 10,000 rpm. Transfer the supernatant to a new Eppendorf tube without disturbing the cell debris.
6. Add 0.5 volumes of 100% ethanol and mix by gentle shaking. Transfer the sample to an RNeasy spin column with 2ml collection tube and centrifuge for 15 sec at 10,000 rpm.
7. Discard the flow through, in order to wash the spin column membrane, add 700 RW1 buffer to RNeasy spin column and centrifuge for 15 sec at 10,000 rpm.
8. Discard the flow through and add 500 µl RPE buffer to spin column and centrifuge for 2 min at 10,000 rpm. Repeat the step.
9. Finally, place the RNeasy spin column into a new collection tube and add 30-50 µl RNase free water directly onto the centre of the column membrane and centrifuge for 1 min at 10,000 rpm to elute the RNA.

DNase Treatment

1. Add the following components in the sterile, nuclease-free tube on ice in the order: water, RNA, buffer, DNase.

S.No.	Reaction Component	Volume
1.	RNA (1-10 ug)	x µl
2.	10 x DNase buffer	5 µl
3.	RNase free DNase	1 µl
4.	Nuclease-free water	x µl (to bring volume up to 50 µl total)

2. Incubate at 37°C for 20 - 30 min.

3. Add 100 µl phenol.

4. Vortex and spin at 14,000 rpm for 10 min at 4°C.

5. Transfer upper phase to a fresh tube.

6. Add 100 µl Chloroform, vortex, spin at 13,000 rpm for 2 min.

7. Transfer upper phase to a fresh tube.

8. Add 3M sodium acetate (pH 5.2) equal to $1/10^{th}$ the volume of the RNA sample.

9. Add 2.5 volumes of 100% ethanol and incubate at -20°C for 1.5 hours.

10. Centrifuge at maximum speed for 10 min at 4°C.

11. Wash pellet carefully with 70% EtOH (-20°C), let air-dry.

12. Re-suspend in 50-100 µl RNase free water

Storage of RNA

Purified RNA may be stored at -20°C or -80°C in RNase free water.

Determining RNA Quality and Quantity

RNA quality can be determined by examining the ratio of absorption at 260 nm and 280 nm with UV spectrophotometry. For high-quality RNA, the A260/A280 ratio should be in the range of 1.9–2.1. RNA can be quantified by measuring the absorption at 260 nm, where 1 absorbance unit is equal to 40µg/ml. In addition, the quality of total RNA preparations should be examined through electrophoresis, where both 18S and 28S RNA bands should be very prominent.

Formaldehyde Agarose Gel Electrophoresis

The overall quality of an RNA preparation can also be assessed by electrophoresis on a denaturing agarose gel. This will give an indication of RNA yield. A denaturing gel system is suggested because most RNA forms extensive secondary and tertiary structures *via* intramolecular base pairing and this prevents it forms migrating strictly according to its size.

Materials Required

MOPS (3-[*N*-morpholino] propanesulfonic acid) 10X buffer, agarose, formaldehyde, DEPC treated water, 2X RNA loading dye, ethidium bromide.

Preparation of Formaldehyde Agarose Gel

To prepare 1% formaldehyde agarose gel, the required quantity of each chemical to be used for different volumes of gel are given below:

Agarose(gm)	0.25	0.50	1
Formaldehyde(ml)	1.35	2.70	5.40
DEPC water(ml)	21.15	42.3	84.6
MOPS 10X (ml)	2.5	5	10
Total Volume (ml)	25	50	100

- Add the agarose in DEPC treated water.
- Heat the mixture to melt Agarose and allow cooling.
- Add formaldehyde, MOPS buffer and ethidium bromide.
- Mix thoroughly and pour onto the gel-casting tray.

Preparation of RNA Samples for Electrophoresis [3]

- Add an equal volume of 2X RNA loading dye to RNA sample and mix well.
- Heat the mixture at 70°C for 10 min.
- Chill on ice and spin down prior to loading on gel.

CONSENT FOR PUBLICATION

Not applicable.

CONFLICT OF INTEREST

The author confirms that this chapter contents have no conflict of interest.

ACKNOWLEDGEMENTS

Declare none.

REFERENCES

[1]   Chomczynski P, Sacchi N. Single-step method of RNA isolation by acid guanidinium thiocyanate-phenol-chloroform extraction. Anal Biochem 1987; 162(1): 156-9.
        [http://dx.doi.org/10.1016/0003-2697(87)90021-2] [PMID: 2440339]

[2]   Chang S, Puryear J, Cairney J. A simple and efficient method for isolating RNA from pine trees. Plant Mol Biol Report 1993; 11(2): 113-6.
        [http://dx.doi.org/10.1007/BF02670468]

[3]   Yu L, Meng Y, Shao C, Kahrizi D. Are ta-siRNAs only originated from the cleavage site of miRNA on its target RNAs and phased in 21-nt increments? Gene 2015; 569(1): 127-35.
        [http://dx.doi.org/10.1016/j.gene.2015.05.059] [PMID: 26026904]

CHAPTER 3

Analyzing Gene Expression through Real Time PCR while Neo-tissue Regeneration using Developed Tissue Constructs

Divakar Singh[1], Tarun Minocha[2], Satyavrat Tripathi[1], Rupika Sinha[1], Shubhankar Anand[1], Hareram Birla[3], Vivek Kumar Pandey[4], Arun Rawat[3], Smita Gupta[3], Sanjeev Kumar Yadav[2], Pawan Kumar Dubey[4] and Pradeep Srivastava[1,*]

[1] *School of Biochemical Engineering, Indian Institute of Technology, Banaras Hindu University, Varanasi-221005, India*

[2] *Department of Zoology, Institute of Science, Banaras Hindu University, Varanasi-221005, India*

[3] *Department of Biochemistry, Institute of Science, Banaras Hindu University, Varanasi-221005, India*

[4] *Centre for Genetic Disorder, Institute of Science, Banaras Hindu University, Varanasi-221005, India*

Abstract: Real-time PCR offers a wide area of application to analyze the role of gene activity in various biological aspects at the molecular level with higher specificity, sensitivity and the potential to troubleshoot with post-PCR processing and difficulties. With the recent advancement in the development of functional tissue graft for the regeneration of damaged/diseased tissue, it is effective to analyze the cell behaviour and differentiation over tissue construct toward specific lineage through analyzing the expression of an array of specific genes. With the ability to collect data in the exponential phase, the application of Real-Time PCR has been expanded into various fields such as tissue engineering ranging from absolute quantification of gene expression to determine neo-tissue regeneration and its maturation. In addition to its usage as a research tool, numerous advancements in molecular diagnostics have been achieved, including microbial quantification, determination of gene dose and cancer research. Also, in order to consistently quantify mRNA levels, Northern blotting and *in situ* hybridization (ISH) methods are less preferred due to low sensitivity, poor precision in detecting gene expression at a low level. An amplification step is thus frequently required to quantify mRNA amounts from engineered tissues of limited size. When analyzing tissue-engineered constructs or studying biomaterials–cells interactions, it is pertinent to quantify the performance of such constructs in terms of extracellular matrix formation while *in vitro* and *in vivo* examination, provide clues regarding the performance of various tissue constructs at the molecular level. In this

** **Corresponding author Pradeep Srivastava:** School of Biochemical Engineering, Indian Institute of Technology (Banaras Hindu University), Varanasi- 221005, India; E-mail: pksrivastava.bce@itbhu.ac.in

Sandeep Kumar & Dhiraj Kumar (Eds.)

chapter, our focus is on Basics of qPCR, an overview of technical aspects of Real-time PCR; recent Protocol used in the lab, primer designing, detection methods and troubleshooting of the experimental problems.

Keywords: Biomaterial, Coding DNA sequences (CDS), DNA, FASTA, Fluorescence, Hybridization, Molecular diagnostics, Master mix, RNA, Real Time-PCR, SYBER green, Tissue Engineering, Thermostability, TaqMan, Wound healing.

INTRODUCTION

There are a number of means to investigate cellular alterations induced by artificial or natural agents during a biological process. Variations in the transcriptome in response to internal or external stimulus may point to proteomic changes. Qualitative and quantitative analysis of pathogens present may reveal significant information to design a treatment regimen. Differential expression of a transgene or inhibition of an endogenous gene can be monitored. In all the objectives, quantitative RT-PCR technique can be employed to cognize the possible outcomes. Its arrival has dramatically changed the field of measuring gene expression. RT- PCR is one of the most widely and extensively used scientific tools for quantitative nucleic acids analysis. With the help of this technique, data processing throughout the whole process becomes quite easier as it allows the detection of a PCR product during exponential phase and amalgamates the process of amplification and detection into one simple step. Basically, RT-PCR technique is the refinement of native Polymerase Chain Reaction (PCR) formulated by Kary Mullis in the 1985 for which he was awarded the Nobel Prize in Chemistry in 1993 for his pioneering work [1]. The previous PCR technique has certain limitations as it involved first amplification, then product analysis, making quantification exceedingly difficult as it gave essentially the same amount of the product independently of the initial amount of DNA template [1]. The first display of RT-PCR was performed with the ethidium bromide (EtBr) dye that fluoresces under ultraviolet light and monitored with a video camera. The combination of UV fluorescence and visualizing the change with a video camera gave birth to the revolutionary real time PCR which quickly matured with time and technology and emerged out as one of the most widespread and influential techniques thereafter [2]. Like any other PCR technique,this also amplifies any essential nucleic acid sequence present in an experimental sample to generate a large number of identical copies which can readily be analysed. But in RT-PCR, the amount of product formed is monitored during the course of the reaction by monitoring the increase in fluorescence of dyes or probes introduced into the reaction that is proportional to the amount of product formed, and the number of amplification cycles required to obtain a particular amount of DNA molecule is registered [3].With significant advancement in software based analysis, optimization and development of highly sensitive detection dye, the amplified nucleotide sequence of the desired sequence in the sample can be easily determined.

In this book chapter, we discuss the principle of RT-PCR, methodology and its application in current research for quantification of gene expression profiles in experimental samples.

BASICS OF RT-PCR

RT-PCR is a versatile and rapid gene expression technique that allows the precise quantification of DNA, cDNA and RNA targets in which the amplified product is detected as a fluorescent signal and measured during each cycle [4]. The intensity of fluorescence is directly proportional to the amount of PCR product that means, the higher the fluorescence, the more the amplified product. For the detection of unlimited supply of amplified DNA molecules either fluorescent dyes or fluorescently labelled sequence-specific probes such as TaqManprobe are used [5].Various things are required for the amplification process such as DNA template (single or double-stranded), two oligonucleotide primers that flank the target sequence to be amplified, dNTPs, a heat-stable polymerase, and magnesium ions in the buffer. PCR instrument is also evoked as a thermal cycler or DNA amplifier and as the name suggests the whole amplification process is performed under desirable temperature conditions (95°C, 50°C and 72°C). The high temperature is required to separate the both the strands of the dsDNA, then the temperature is reduced to its annealing temperature to anneal the primers to the template and finally, the temperature is maintained around 72°C, which is optimum for the polymerase enzyme that extends the primers by incorporating the dNTPs [6]. Also the melting temperature should be optimum, mostly in between 50-60°C (depends on the length and sequence of the template) to fully separate the strands of the template as partially separated strands will rapidly reanneal when temperature decreases and there will be no priming [7]. PCR is one of the most extensively used technique in different facets of Life sciences. With its help, specific sequences amplify many million folds using sequence-specific oligonucleotide (primers), thermostable DNA polymerase and temperature specific cycling. It has got several advantages from traditional PCR, in which detection and quantification of the amplified sequence are performed at the end of the reaction during the last PCR cycle and involve post-PCR analysis such as gel electrophoresis and further image analysis. While in the real-time qPCR, the amplified product is measured with great precision at each cycle. It amplifies DNA exponentially, doubling the number of target molecules with each amplification cycle. The amount of DNA is measured after each cycle *via* fluorescent dyes (or probes) that yield an increasing fluorescent signal which is directly proportional to the number of amplified product molecules formed. Data collected in the exponential phase of the reaction yield quantitative information on the starting quantity of the amplification target. Fluorescent reporters used in real-time PCR include double-stranded DNA (dsDNA) binding dyes or dye molecules attached to PCR primers or probes that hybridize with PCR products during amplification. This change in fluorescence over the course of the reaction is measured by a thermal cycler instrument that combines thermal cycling with fluorescent dye scanning capability. By plotting fluorescence against the number of cycles, the thermal cycle generates an amplification plot (Fig. **1**) that represents the accumulation of the product over the duration of the entire PCR.

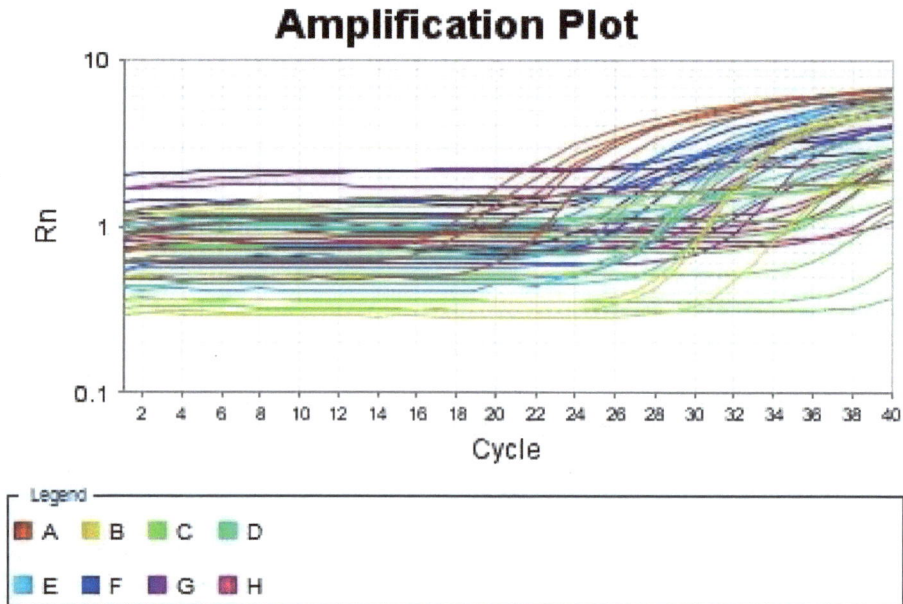

Fig. (1). Amplification plot generated by qPCR machine. Different colours represent expression pattern of genes used in study. As number of cycles are increased the plot of all genes increases respectively.

Merits of RT-PCR

There are various advantages of RT-PCR which are as follows:

- The primary advantage of RT-PCR is the ability to detect and identify amplified fragments during the whole process in a single tube surpassing post-PCR manipulations.
- Ability to precisely measure the amount of amplicon in each cycle, which allows highly accurate quantification of the amount of starting material in samples.

- In RT-PCR, the amount of the product is reckoned during the exponential phase while in case of standard PCR the product is reckoned during the plateau phase.
- With the use of RT-PCR, a large number of copies of the specific DNA segment are produced instantly.

PCR Cycling

The PCR is a chain reaction because newly synthesized DNA strands will act as templates for further DNA synthesis in subsequent cycles. Theoretically, PCR generates thousands to millions of copies of a particular section of DNA from a very small amount of DNA [8]. In each stage of the cycle proper optimization of time and temperature is required for the combination of template and primer. In most cases, reactions are generally run for 40 cycles which is sufficient to differentiate an amplified product with respect to endogenous control taken (Fig. 2).Various endogenous controls such as 18S rRNA, β-actin, GAPDH, β-tubulin, phosphoglycerate kinase1 and TATA box binding protein *etc.* are commonly used for relative quantification [9].

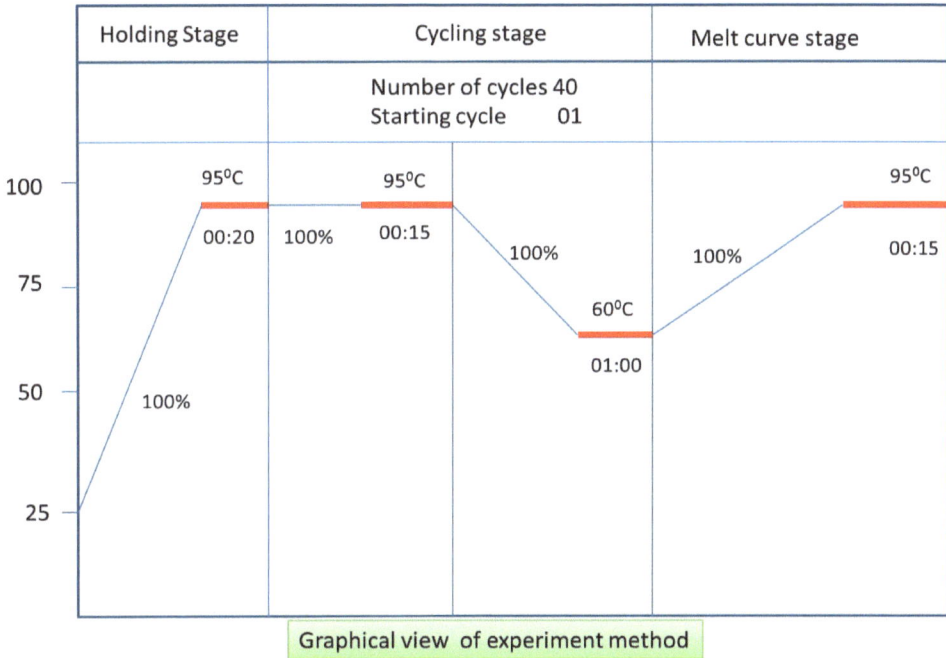

Fig. (2). Diagrammatic representation of qPCR steps.

Generally, the PCR reaction cycle comprises following successive steps which are as follows:

1. **Initial Denaturation:** This is the first and formost step in the PCR where temperature is enhanced to 92-95°C and incubated for 2–5 min (depending on enzyme characteristics and template complexity) to ensure that all dsDNA molecules are isolated into single strands. It is important that the temperature is maintained at this stage for long enough to ensure that the DNA strands have separated completely but not so long that the DNA gets damaged. If DNA gets damaged during the initial denaturation step it results in decreased detection sensitivity.

2. **Denaturation:** This shorter denaturation step (10 sec to 1 min at 95 °C) is used for high-temperature incubation to "melt" dsDNA by disrupting the hydrogen bonds between the bases in two strands of template DNA (complementary bases). This results in two single strands of DNA, which will act as templates for the production of the new strands of DNA. The denaturation time can be enhanced if the GC content is high enough in the template DNA.

3. **Primer Annealing:** During this step, the temperature is lowered to approximately 5°C below the melting temperature (Tm) of the primers (30 sec to 1 min, at 45–60°C) to enable the DNA primers to attach to the template DNA. Once the Primers gets attached to the template the polymerase initiates the process of DNA synthesis. During annealing, complementary sequences hybridize with each other.

4. **DNA Synthesis or Extension:** During this final step, the temperature is increased typically to 72°C to synthesize a new DNA strand complementary to the DNA template strand by a special polymerase enzyme known as Taq DNA Polymerase which adds dNTPs from the reaction mixture one by one. This step lasts for 20 sec to 1 min at 70-72°C because the activity of the DNA polymerase is optimal, and primer extension occurs at the rate of 100 bases per sec. It attaches to the primer and then adds DNA bases to the single strand one-by-one in the 5' to 3' direction.

5. **Repeat:** Denaturation, annealing and extension are the steps constitute one cycle of PCR amplification and repeated in a cyclic manner, resulting in huge number of copies of the specific DNA segment just by changing the temperature of reaction mixture. All above mentioned cycles are based on heating and cooling reaction. Each reaction comprises of definite time-dependent controlled temperature incubation. In case the sample is different or more detection sensitivity is required, these conditions are optimized accordingly.

RT-PCR Master Mix Components

RT-PCR Master Mix (SYBR green, Light cycler SYBR green) is a complete formulation that contains all the necessary ingredients required for successful PCR amplification in single tube *i.e.,* TaqMan DNA Polymerase, RNase Inhibitor, dNTPs, magnesium ions and requisite buffer. There is no requirement of adding all components separately like as in conventional PCR. Simply this master mix is added to nucleotide template, primers, RNase-free water and the reaction is ready. During RT-PCR,thermostable DNA polymerase first synthesizes sec-strand cDNA to the first-strand cDNA molecule. This creates a double-stranded DNA template which is exponentially amplified in subsequent rounds of thermal cycling.

Thermostable DNA Polymerase

Thermostable DNA polymerase is one of the crucial players that affect the PCR specificity. If there is nonspecific primer binding to Cdnatemplate, it allows the polymerase to synthesize anon-specific product. That's why to tackle this problem a "hot-start" enzyme method is used which deactivates DNA either reaction setup or denaturation step.

Deoxyribonucleotidetriphosphates (dNTPs)

dNTPsare also an integral part of thermal cycle reaction. They are basically a blend of four monomeric units of each nucleotide namely, dATP, dTTP, dCTP and dGTP, which serve as essential "building blocks" for anew strand of DNA strands.

Divalent Cation Especially Magnesium Ions

In RT-PCR, magnesium ions play a crucial role and are available either as chloride or magnesium sulphate. Magnesium ions act as a cofactor for Taq polymerase enzyme. During the PCR reaction when dNTP's disassociates into dNMP's to form a phosphodiester bond between 3' OH of adjacent nucleotide and 5' phosphate of the forthcoming nucleotide the Mg^{+2} bind to the alpha (α) phosphate group of dNTP and removes the beta (β) and gamma (γ) phosphate from dNTP.

Template

The DNA template is that particular DNA sequence which we want to copied. Nearly 1μg of cDNA generated from ~ 100ng of total RNA is needed for proper template during PCR reaction. RNA should be devoid of any RNase contamination and sterile conditions should be maintained. Necessary measures should be taken on quality and quantity of template taken as any contamination would lead to multiple problems in amplification. Thus high-quality cDNA synthesis is important for accurate gene expression analysis.

RT-PCR Primer Design

Appropriate primer designing is the most critical and fundamental aspect for amplification of desired fragment. While designing primers, some prior DNA sequence information is essential for the amplification process. The amplicon length should not be more than 200bp, as longer products do not amplify efficiently. In addition, the length of the primers should not be very short and long, generally, 18–21 nucleotides longprimers should be preferred according to standard PCR guidelines [10]. They should be highly specific and be free of an internal scary structure such as hairpin loop structure. Additionally, if both primers (Left/Right)have drastically different melting temperatures (Tm), it hinders efficient binding of both primers to their target sequence during thermal cycling.

There are certain requirements which must be fulfilled while designing primers for RT-PCR. These re equipments are as follow-

1. Primer length should be 18-21 nucleotides long.
2. GC content should be varied between 50 and 60% with an even distribution of all four nucleotides.
3. Primer pairs should have optimal melting temperatures *i.e.* 55-60°C and the melting temperatures for two primers used together should not differ by > 5°C (If T_m is kept within 1°C it gives primer with better annealing capability).
4. The 3' end of primers should always have either G or C in order to clamp the primer and to prevent "breathing" of ends and to increase priming efficiency.
5. The primer pair sequences should be deeply analysed to avoid complementarities and hybridization of primer-dimer formation.
6. Single bases such as TTTT or di-nucleotide recurrences such as GCGCGC should be avoided as they can hybridize to non-target sites and may lead to the formation of hairpin loop structures that ultimately provide undesired amplification product.
7. At last, the primers must be cross checked for their specificity. This can be achieved with the help of various tools and databases available online such as NCBI Primer-Blast (https://www.ncbi.nlm.nih.gov/tools/primer-blast/). With these tools, designed primers can be checked whether they are specific and recognize the desired target site.

Primer Designing

As mentioned earlier the primers are integral parts of this molecular biology techniques. Without proper designing of appropriate primer, none of above steps can be performed in thermal cycler. Therefore, extra care should be taken while designing of the primers. For this various software programs, such as Primer3 Software accessed by the broad institute (http://bioinfo.ut.ee/primer3/) can be used which is based on mathematical algorithms thus automatically evaluating the target sequence entered by the user and design primers according to the instructions provided by the user. The biggest benefit while using such software is that they enable users to generate primers that are highly specific for the target sequences and devoid of scary structure and avoid complementary hybridization at 3′ ends within each primer and with each other at no extra cost.

Basics of Primer Designing

There are certain steps which can help any user to design primer for experimental studies. These steps are as follows-

1. Firstly search for the protein sequence of a target gene using NCBI database (https://www.ncbi.nlm.nih.gov/pubmed/).
2. Click on that searched gene query and go for its coding DNA sequences (CDS). CDS are the regions of a gene composed of exons, that code for protein.
3. Clicking on CDS will display text terms like FASTA, Gene bank and help.
4. Now click on FASTA which will take user towards complete nucleotide sequence of that protein.
5. Now open primer3 tool provided by the broad institute (http://bioinfo. ut.ee/primer3/) and paste complete nucleotide sequence provided in FASTA format.
6. Also enter essential details like primer size (*e.g.* 18-21 nucleotides), primer Tm (*e.g.* 55-60°C), primer GC content (*e.g.* 50-60%) and product size range (*e.g.* 100-200 nucleotides).
7. Now click on Pick primers which will open a separate window displaying a list of primers generated according to the entered information.
8. At last check the primers for their specificity by using the Primer-Blast tool provided by NCBI (https://www.ncbi.nlm.nih.gov/tools/primer-blast/). Simply enter details like forward/reverse primer and organism (*e.g.Musmusculus*) against which the primers are being developed.

RT-PCR Fluorescence Detection Chemistry

Over the last few decade, significant research has been flourished for the development of such methods which enable a highly sensitive method for DNA detection in qPCR. Basically all these methods have been employed into two major sets based on the fluorescent reporter (SYBR green, LC green, EVA green and BOXTO dye) and the specificity (fluorophores attached to oligonucleotide) of thePCR detection [11, 12]. Among both, these methods detection through SYBR green and TaqMan probes are very common these days whose mechanism and detection are described further.

SYBR Green-based Detection

SYBR green is a green fluorescence DNA binding dye commercially available in the market. Chemically SYBR Green is an asymmetrical cyanine dye (N', N'-dimethyl-N-[4-[(E)-(3-methyl1,3-benzothiazole-2-ylidene)-ethyl]-1-phenylquinolin-1-ium-2-yl]-Npropylpropane-1, 3-diamine) having high binding affinity for dsDNA [13]. The DNA-dye interaction absorbs blue light at 497 nm and emits fluorescence (green light)at 520 nm. Although the dye preferentially binds to the dsDNA but it could also stain ssDNA and RNA with lower performance [14].

Mechanism of Action

The fluorescent dye SYBR Green has the property to intercalates non-specifically to the minor groove of the dsDNA [15]. Before commencing the amplification, the reaction mixture contains the denatured template, the primers, and the master mix (2X) containing dye. The unbound dye depicts very little fluorescence; However, the intensity of fluorescence is substantially enhanced when the dye is bound to DNA. After annealing of primers dye molecules starts binding to the double strand resulting an increase in fluorescent signal. Further, as long as the elongation continues more dye molecules bind to the newly synthesized dsDNA (Fig. **3**). The instrument continuously displays an increase in the intensity of fluorescence in RT-PCR. During this whole process nonspecific products or primer-dimerization may also takes place hence, a melting curve analysis is suggested to check the specificity of the amplicon [16].

Advantages

- The costs of fluorescent dyes such as SYBR green are much lower than those of fluorescent probes.
- No probe is required, which can reduce assay setup and running costs.
- It can be used to monitor the amplification and detection of any dsDNA sequence.

Limitations

- SYBR green monitors the total amount of dsDNA, but can't distinguish different sequences.
- SYBR green has however, a limitation that includes preferential binding to G.C rich sequences.
- The biggest disadvantage using SYBR green is that it is nonspecific and will

bind to any amplified dsDNA thus interfering with gene expression results. However, this problem canbe tackled out by checking the specificity by speculating a melt curve analysis at the end of the PCR usually present in all currently available RT-PCR instruments.

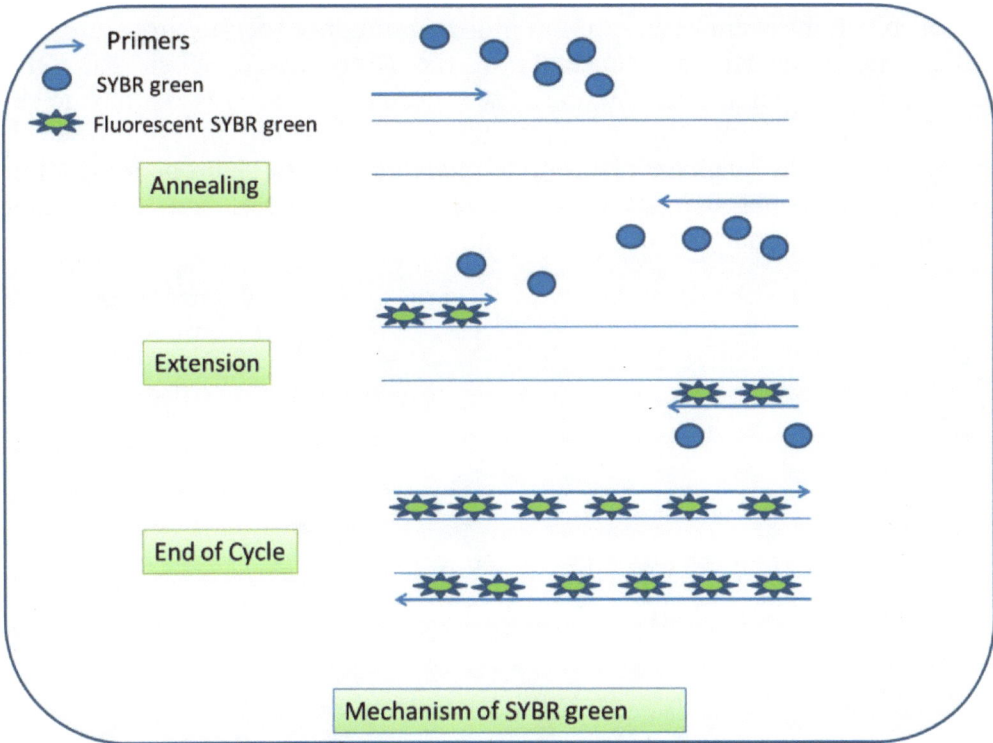

Fig. (3). Schematic representation of action of SYBR green. As there is increase in amount of DNA during each. amplification cycle more SYBR green gets bound to it corresponding to its increase in fluorescence signal.

TAQMAN probe Based Detection

To be sure that the correct target sequence is being amplified, a sequence specific fluorescent probe is required. TaqMan probe is the finest instance which is oligonucleotide designed to hybridize to an internal region of the PCR product. TaqMan probes are highly sensitive hydrolysis probes which increases the specificity of qPCR [17].

Mechanism of Action

The TaqMan probes simply rely on the 5′–3′ exonuclease activity of Taq DNA Polymerase to cleave them during hybridization to the complementary target-template sequence [18]. The TaqMan probe has three elements a short wavelength fluorophore covalently linked to the one end *i.e.* 5' end, a sequence that is specific for the target DNA and a quencher at the other end *i.e.* 3'end (Fig. **4**). The quencher molecule present at 3' end quenches the fluorescence emitted by the fluorophores when excited by the instrument's light source. As long as both the fluorophores and the quencher are in intimacy, fluorescence is quenched and no fluorescence light is emitted. TaqMan probes are highly specific in nature, they are designed to anneal to the center of the target DNA. When Taq DNA polymerase starts polymerization, the 5' to 3' exonuclease activity of enzyme cuts the probe into single nucleotides. This annuls the close proximity between the fluorophore and the quencher resulting in the release of the fluorophore from the probe, thus abolishes the quenching effect and allowing fluorescence that is proportional to the number of newly synthesized strands.

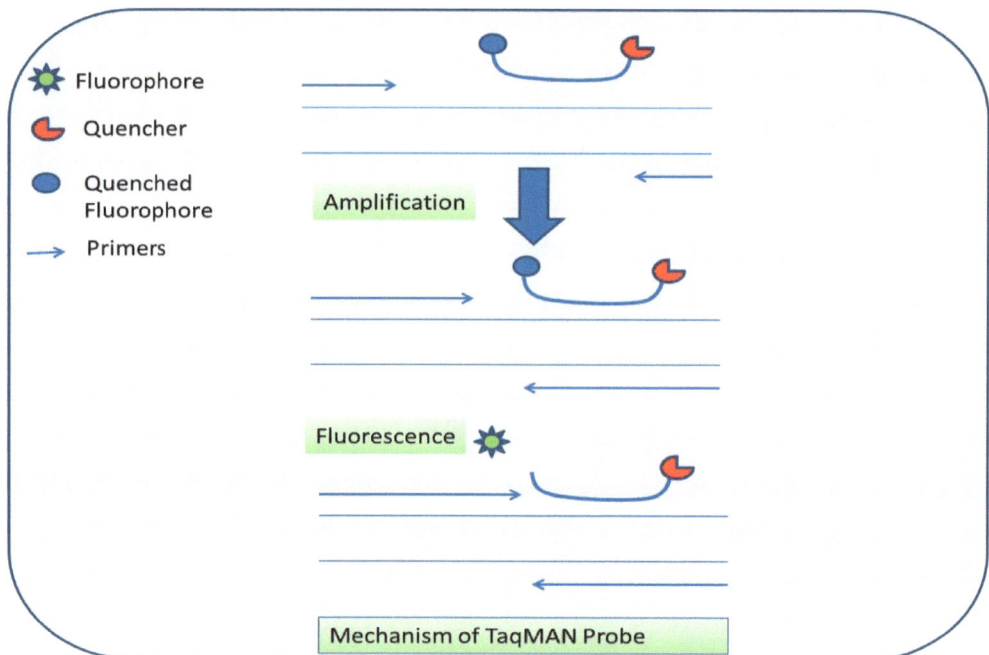

Fig. (4). Schematic representation of a RT-PCR employing TaqMan probe detection assay.

Advantages

- TaqMan probe-based assays are extensively used as a potent tool in qPCR in various laboratories for research purpose.
- The main advantage is that a specific hybridization between probe and target is required to generate fluorescent signal.
- Probes can be labelled with different reporter dyes which allows the amplification and detection of multiple target sequences in a single reaction tube

(multiplex Real-Time PCR) can be performed.

Limitations

- The primary disadvantage is that for different sequences synthesis of different probes is essential.
- Other limitation is that it requires precise assay setup and is very expensive.

RT-PCR Experimental Design and Gene Expression Profiling

Though transcription of mRNA is quite sensitive but any mistake during reaction setup can alter the experiment results. Therefore, any experimental design (Fig. **5**) is the key to any gene expression study which should be performed under well-defined conditions. Each and every parameter such as experimental conditions, proper handling, control groups, type and number of replicates should be stringently monitored prior to conducting RT- PCR experiments to ensure good biological reproducibility.

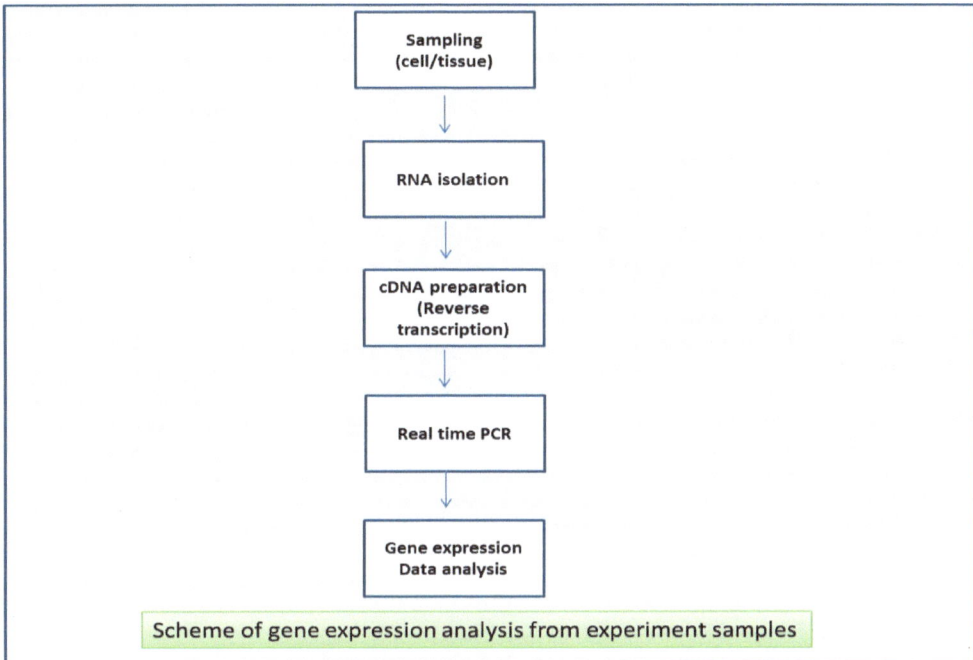

**Sampling
(cell/tissue)**

↓

RNA isolation

↓

**cDNA preparation
(Reverse
transcription)**

↓

Real time PCR

↓

**Gene expression
Data analysis**

Scheme of gene expression analysis from experiment samples

Fig. (5). Overview of gene expression profiling using qPCR.

To analyse the expression pattern of various genes the sample should be homogenized thoroughly in order to increase the RNA yield. The samples which are collected for the further analysis should be used to obtain the RNA samples.

While isolating RNA extra precaution should be given in order to avoid the contamination of RNases in sample solutions, consumables, and lab wares. DEPC treatment is enough to inactivate the RNases on lab wares and other sites. Further, RNA samples should also be treated with DNases in order to avoid genomic DNA contamination. After the RNA has been extracted from samples its quantity and purity should be determined by measuring the ratio of the UV absorbance at 260 nm and 280 nm using a nanodrop or Pico-drop. After this step the extracted RNA is used for preparation of cDNA which will be used in further real time gene expression studies. In addition, the selection of reference gene is also a crucial factor while performing the experiment. In RT-PCR experiments, reference genes are used as control to normalize the data by correcting the differences in quantities of cDNA used as a template. A reference gene is one which expresses continuously in any experimental conditions or time point in each and every cell of the tissues. Several reference genes such as GADPH, Beta-actin, 16S-rRNA are commonly used as reference genes. At last when all these essential requirements have been fulfilled the gene expression levels are calculated from the generated data. A sample experimental layout of experiment has been shown in Fig. (**6**). A RT-PCR experiment generates a Cycle Threshold (C_T) value for each gene in each experiment sample which reflects the transcriptional activity of that gene in the experimental sample. After selecting the genes to be analysed by RT-PCR, the expression profiling by this technique has many advantages. It generates high-quality data with higher sensitivity and since all genes expressions are studied under same conditions. Further, one can perform multiple repeats and speculate the samples with RT-PCR, a prominent requirement for the statistical analysis of data.

Setting up a Reaction Mixture

The preparation of the reaction mixture solely depends on the concentration of the stock reagents. The final concentration is prepared according to the desired conditions. For one reaction volume per well, the respective volume of following components is added to make up the final volume of 10µl. The list of all components given in the below table:

Final concentration of Master mix= 2X Final concentration of Forward Primer = 100nM Final concentration of Reverse Primer= 100Nm	
Components	**Volume (µl) of 1 Reaction**
Master mix (2x)	5 µl
Forward primer	0.5 µl
Reverse primer	0.5 µl

Cont.....

Sample	1 μl
Water	3 μl
Total volume	10 μl

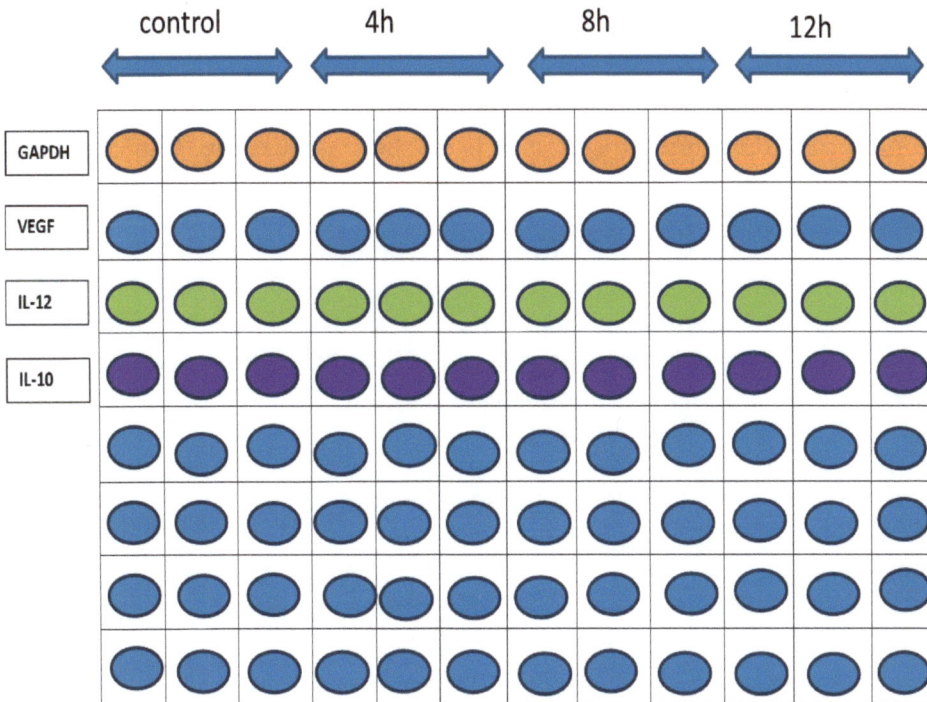

Fig. (6). Diagrammatic representation of plate setup during real time PCR. Different colours represent various genes whose expression analysis would be done during study.

Basic PCR Protocol and Establishment of Reaction Cycle

RT-PCR instrument expeditiously heats and cools the reaction mixture, allowing the denaturation of dsDNA for strand separation and allowing primers to anneal towards both the strands of the DNA template and lastly the elongation of the PCR product. The number of cycles can be adjusted accordingly, although 40 cycles are preferred.

In general, the PCR mechanism commences with the denaturation step that takes place at 92-95°C depending on the required temperature for DNA polymerase activity and GC content of the template DNA. Basically,one min denaturation at 94 °C is given before the start of the reaction. If the duration is increased, it may result in inactivation of the DNA polymerase activity. In the next step, the thermal cycler initiates different rounds of a three-step temperature cycle. The three temperature steps in a single cycle accomplish three tasks: the first step denatures the template (and in later cycles, the amplicon as well), the sec step allows optimal annealing of primers, and the third step permits the DNA polymerase to bind to the DNA template and synthesize the PCR product. The duration and temperature of each step within a cycle may be altered to optimize production of the desired amplicon. Nearly 40 cycles are sufficient to form sufficient PCR product, above this, if more cycles are done, it may result in the formation of undesirable products.

The time for the denaturation step is kept as short as possible. Usually, 10 to 60 secs is sufficient for most DNA templates. The temperature for this step is usually the same as that used for the initial denaturation phase (*e.g.* , 95°C). A 30-sec annealing step follows within the cycle at a temperature set about 5°C below the apparent Tm of the primers (ideally between 50°C to 60°C). Further, the cycle concludes with an elongation step. The temperature (usually 70°C to 80°C) depends on the DNA polymerase selected for the experiment. The final phase of thermal cycling incorporates an extended elongation period of 5 min or longer. This last step allows the synthesis of many uncompleted amplicons to finish and, in the case of Taq DNA polymerase, permits the addition of an adenine residue to the 3' ends of all PCR products. This modification is mediated by the terminal transferase activity of Taq DNA polymerase and is useful for subsequent molecular cloning procedures that require a 3'-overhang. Finally,the reaction is put toan end by chilling the mixture to 4°C. Following steps are being performed for gene expression analysis:

1. Keep a 96 well plate into the ice box.
2. Pipette the following reaction components in the mentioned order in PCR tubes: Master Mix, Nuclease-free water, Water, Forward and reverse primer and template DNA.
3. Switch on the thermal cycler and design the experiment layout and tag all genes and samples to be studied.
4. Set the desired command in Real-time PCR machine such as temperature and cycle number.
5. Recheck all the amplification stings and experimental settings.
6. At last, run the machine for amplification of the desired sequence.
7. After completion of all cycles simply export the data and do data analysis to find out whether the target genes have been up-regulated or down-regulated.

Application of RT-PCR

Over the last few years, with the improvement and refinement of PCR protocols and due to the availability of commercial automatic thermal cyclers, the applications of PCR have been increased manifold. Now a days, RT-PCR technology has applications in different facets of life sciences including molecular biology, genetics, microbiology, diagnostics, clinical laboratories, forensic science, environmental science, food technology and many other disciplines of life sciences. RT-PCR is most widely and extensively used as a potent tool for gene amplification in various laboratories for research purpose. As far as molecular biology area is concerned, PCR is used in DNA cloning and sequencing, Southern blotting and recombinant DNA technology. In clinical microbiology laboratories, PCR is pivotal tool for the diagnosis of microbial infections and epidemiological studies. In food and dairy sciences, PCR acts as a crucial player for the agricultural and food industries as an incredible method as compared to the traditional ones. When forensic area is taken into account, PCR is very reliable because of only scarce amount of DNA is required for instance, sufficient amount of DNA can be attained from a tiny droplet of blood or a single hair. Hence, application of RT-PCR in combination with several other techniques have trigerred the monitoring of both therapeutic intervention and individual responses to the various drugs. The introduction of these new methods in the fields of human practices induced rapid expansion of molecular approaches, advances in diagnostic imaging technology, surgical management, and therapeutic modalities. Currently, it is the most sensitive method to quantify the specific DNA to detect even a single molecule and diagnostics become feasible with lower amounts of complex biological materials compared to traditional methods [19]. Cancer is the second leading cause of mortality across the globe and due to this, research has been well documented in cancer field [20, 21]. Most of the prevalent cancers have been detected by measuring the expression of either marker genes or probes. The sensitivity of single marker assays is not high enough for clinical applications [22] hence multidimensional approach is necessary. Adopting a multigene panel for most common malignant diseases such as breast cancer, cervical cancer, colorectal cancer, prostate cancer not only increased the accuracy of diagnosis of various diseases but also the development of effective therapeutic strategies [23 - 25]. RT-PCR is a reliable tool to monitor the progress and report of cancer therapeutic agents that can be of immense use. Recently numerous kits are commercially available for clinical tests, and these developments uplifted the use of RT PCR in several other fields of human practices [26].

CHALLENGES

Researchers have concentrated first to study the involvement and impact of various genes and then how they regulate different biological functions has been investigated. This investigation has followed a progressive sense in a way that first discovery then prediction of the target, then function to systems perspective and further organism perspective were taken into account. The investigation deals with certain points such as how different cell types are formed from a genome, the contribution of the genome in the basic and specialized functions of the cells and in what ways the cell interacts with its environment. To understand and clarify such objectives, RT-PCR has emerged as an appreciable methodical tool which depicts the functional connections between coding and noncoding regions of human genome. Interactions between multiple genes are responsible for affecting heritable traits, including disease susceptibility. Though, the understanding about how the genes interact is less yet, very few possible genetic interactions have been observed experimentally. Plenty of data attained through RT PCR for understanding the functional expression of genes can be analysed easily by the concept, that genes and non-genes comprise fractal sets, determining the ensuing fractal hierarchies of complexity of biological processes. To achieve this goal, the sequencing of the genomes is required for a number of model organisms that can help to provide a strong framework. The methods used for gene expression profiling and protein interaction mapping in order to identify pathways involved in progression of disease are largely being used.

CONSENT FOR PUBLICATION

Not applicable.

CONFLICT OF INTEREST

The author confirms that this chapter contents have no conflict of interest.

ACKNOWLEDGEMENTS

Declare none

REFERENCES

[1]   Kubista M, Andrade JM, Bengtsson M, *et al.* The real-time polymerase chain reaction. Mol Aspects Med 2006; 27(2-3): 95-125.
[http://dx.doi.org/10.1016/j.mam.2005.12.007] [PMID: 16460794]

[2]   Abrantes LB. Validação de genes alvo da *via* Rac1/PAK1-BCL6/STAT5 envolvidos na progressão tumoral (Doctoral dissertation). 2016.

[3]   Bustin SA. Absolute quantification of mRNA using real-time reverse transcription polymerase chain reaction assays. J Mol Endocrinol 2000; 25(2): 169-93.
[http://dx.doi.org/10.1677/jme.0.0250169] [PMID: 11013345]

[4]   Wong ML, Medrano JF. Real-time PCR for mRNA quantitation. Biotechniques 2005; 39(1): 75-85.
[http://dx.doi.org/10.2144/05391RV01] [PMID: 16060372]

[5]   Tajadini M, Panjehpour M, Javanmard SH. Comparison of SYBR Green and TaqMan methods in quantitative real-time polymerase chain reaction analysis of four adenosine receptor subtypes. Adv Biomed Res 2014; 3: 85.
[http://dx.doi.org/10.4103/2277-9175.127998] [PMID: 24761393]

[6]   Lorenz TC. Polymerase chain reaction: basic protocol plus troubleshooting and optimization strategies. JoVE (Journal of Visualized Experiments) 2012; 1(63): e3998.
[http://dx.doi.org/10.3791/3998]

[7]   Ibrahim Z, Kurniawan TB, Khalid M, Eds. A DNA sequence design for molecular computation of HPP with output visualization based on Real-Time PCR. 2007 IEEE Congress on Evolutionary Computation. In: IEEE; 2007.
[http://dx.doi.org/10.1109/CEC.2007.4424694]

[8]   Freeman WM, Walker SJ, Vrana KE. Quantitative RT-PCR: pitfalls and potential. Biotechniques 1999; 26(1): 112-122, 124-125.
[http://dx.doi.org/10.2144/99261rv01] [PMID: 9894600]

[9]   Kozera B, Rapacz M. Reference genes in real-time PCR. J Appl Genet 2013; 54(4): 391-406.
[http://dx.doi.org/10.1007/s13353-013-0173-x] [PMID: 24078518]

[10]   Abd-Elsalam KA. Bioinformatic tools and guideline for PCR primer design. Africa J Biotechnol 2003; 2(5): 91-5.

[11]   Eischeid AC. SYTO dyes and EvaGreen outperform SYBR Green in real-time PCR. BMC Res Notes 2011; 4(1): 263.
[http://dx.doi.org/10.1186/1756-0500-4-263] [PMID: 21798028]

[12]   Lin L, Li R, Bateman M, Mock R, Kinard G. Development of a multiplex TaqMan real-time RT-PCR assay for simultaneous detection of Asian prunus viruses, plum bark necrosis stem pitting associated virus, and peach latent mosaic viroid. Eur J Plant Pathol 2013; 137(4): 797-804.

[13]   Shu P-Y, Chang S-F, Kuo Y-C, Yueh Y-Y, Chien L-J, Sue C-L. Development of group-and serotype-specific one-step SYBR green I-based real-time reverse transcription-PCR assay for dengue virus. J Clin Microbiol 2003; 41(6)2408-16.

[14]   Giglio S, Monis PT, Saint CP. Demonstration of preferential binding of SYBR Green I to specific DNA fragments in real-time multiplex PCR. Nucleic Acids Res 2003; 31(22)e136.
[http://dx.doi.org/10.1093/nar/gng135] [PMID: 14602929]

[15]   Zipper H, Brunner H, Bernhagen J, Vitzthum F. Investigations on DNA intercalation and surface binding by SYBR Green I, its structure determination and methodological implications. Nucleic Acids Res 2004; 32(12)e103
[http://dx.doi.org/10.1093/nar/gnh101] [PMID: 15249599]

[16]   Navarro E, Serrano-Heras G, Castaño MJ, Solera J. Real-time PCR detection chemistry. Clin Chim Acta 2015; 439: 231-50.
[http://dx.doi.org/10.1016/j.cca.2014.10.017] [PMID: 25451956]

[17]   Nagy A, Vitásková E, Černíková L, *et al.* Evaluation of TaqMan qPCR system integrating two identically labelled hydrolysis probes in single assay. Sci Rep 2017; 7: 41392.
[http://dx.doi.org/10.1038/srep41392] [PMID: 28120891]

[18]   Didenko VV. DNA probes using fluorescence resonance energy transfer (FRET): designs and applications. Biotechniques 2001; 31(5): 1106-1116, 1118, 1120-1121.
[http://dx.doi.org/10.2144/01315rv02] [PMID: 11730017]

[19]   Kalow W. Pharmacogenetics and pharmacogenomics: origin, status, and the hope for personalized medicine. Pharmacogenomics J 2006; 6(3): 162-5.
[http://dx.doi.org/10.1038/sj.tpj.6500361] [PMID: 16415920]

[20]   Alaoui-Jamali MA, Xu YJ. Proteomic technology for biomarker profiling in cancer: an update. J Zhejiang Univ Sci B 2006; 7(6): 411-20.
[http://dx.doi.org/10.1631/jzus.2006.B0411] [PMID: 16625706]

[21]   Liefers GJ, Tollenaar RA. Cancer genetics and their application to individualised medicine. Eur J Cancer 2002; 38(7): 872-9.

[http://dx.doi.org/10.1016/S0959-8049(02)00055-2] [PMID: 11978511]

[22]   Khanna C, Helman LJ. Molecular approaches in pediatric oncology. Annu Rev Med 2006; 57: 83-97.
        [http://dx.doi.org/10.1146/annurev.med.57.121304.131247] [PMID: 16409138]

[23]   Ståhlberg A, Zoric N, Åman P, Kubista M. Quantitative real-time PCR for cancer detection: the
        lymphoma case. Expert Rev Mol Diagn 2005; 5(2): 221-30.
        [http://dx.doi.org/10.1586/14737159.5.2.221] [PMID: 15833051]

[24]   Hoebeeck J, van der Luijt R, Poppe B, *et al.* Rapid detection of VHL exon deletions using real-time
        quantitative PCR. Lab Invest 2005; 85(1): 24-33.
        [http://dx.doi.org/10.1038/labinvest.3700209] [PMID: 15608663]

[25]   Espy MJ, Uhl JR, Sloan LM, *et al.* Real-time PCR in clinical microbiology: applications for routine
        laboratory testing. Clin Microbiol Rev 2006; 19(1): 165-256.
        [http://dx.doi.org/10.1128/CMR.19.1.165-256.2006] [PMID: 16418529]

[26]   Deepak S, Kottapalli K, Rakwal R, *et al.* Real-time PCR: revolutionizing detection and expression
        analysis of genes. Curr Genom 2007; 8(4): 234-51.
        [http://dx.doi.org/10.2174/138920207781386960] [PMID: 18645596]

CHAPTER 4

A Modified Western Blot Protocol for Enhanced Sensitivity in the Detection of a Tissue Protein

Sachchida Nand Rai, Mallikarjuna Rao Gedda, Walia Zahra, Hareram Birla, Saumitra Sen Singh, Payal Singh, Neeraj Tiwari, Rakesh K. Singh and Surya Pratap Singh[*]

Department of Biochemistry, Institute of Science, Banaras Hindu University, Varanasi, India

Abstract: Western blots (WB) are designed to investigate protein levels and their patterns of modification in homogenized tissue samples. Although, Western blots are quantifiable, unlike immunohistochemistry, cellular integrity is lost. The availability of antibodies against the protein and their patterns of modification of interest form the basis of both Western blots and Immunohistochemistry. Antibodies can also be directed not only against proteins but against chemical modifications of the proteins too, such as phosphorylation and glycosylation of specific amino acid residues. In Western blotting, the proteins in the sample are denatured, size-separated on a denaturing acrylamide gel, and transferred to a nylon membrane. Antibody paratopes can then bind to the antigenic epitope in the protein present on the nylon membrane. Thus, with the help of a chemiluminescent assay system that darkens X-ray films, the resulting antibody-antigen complex can be visualized. Because of the ubiquitous and relatively inexpensive availability of WB equipment, the quality of WB in publications and following analysis and investigation of the data can be variable, possibly resulting in forged conclusions. This may be because of the poor laboratory technique and/or lack of understanding of the significant steps involved in WB and what quality control procedures should be followed to ensure effective data generation. The present book chapter focuses on providing a detailed description and critique of WB procedures and technicalities, from sample collection through preparation, blotting, and detection, to examination of the data collected. We aim to provide the reader with the improved expertise to decisively carry out, assess, and troubleshoot the WB process, in order to produce reproducible and reliable blots.

Keywords: Western Blot, Immunohistochemistry, Antibody, Protein, Membrane.

INTRODUCTION

The technique of Western blotting is basically used to quantify the expression level of the proteins and their modification in homogenized tissue samples and ly-

[*] **Corresponding author Surya Pratap Singh:** Department of Biochemistry, Institute of Science, Banaras Hindu University, Varanasi, India; E-mail: suryasinghbhu16@gmail.com

sed cell-lines. Although Western blots (WB) and Immunohistochemistry (IHC) are quantifiable, the only disadvantage of WB over IHC is the cell loses its form and integrity during homogenization and sonication processes. Both WB and IHC depend on the same principle of availability of a specific antibody paratope against the epitope of antigen (protein of interest). During WB, the antibodies which were used directed not only against the protein of interest but also their specific chemical modifications including glycosylation and phosphorylation of the amino acid residues present in them [1].

WB includes the denaturation of the proteins in the homogenized tissue samples and lysed cell-lines which are further subjected to acrylamide gel electrophoresis by which they are separated on the basis of size, and further transferred on to a nylon membrane. The membrane is further incubated with the antibody which reacts with the protein of interest on the membrane by binding to its specific antigen. The resulting Antigen-Antibody complex is generally visualized using chemiluminescent assay system by darkening X-ray films or by using gel documentation system.

SAMPLE PREPARATION

The samples used during WB are prepared commonly by sonication or homogenization of the tissue [2]. Protein extraction technique helps to break the cell so as the cytosolic proteins can be collected. The isolation of the protein is done at a lower temperature and the protein inhibitors are used to prevent the protein from denaturation [3]. A UV-visible spectrophotometer is used to quantify the concentration of the proteins in the sample as shown in Fig. (**1**). By using this concentration, the mass of protein which has to be loaded in the well can be quantified by evaluating the relationship between concentration, mass, and volume. The proteins are diluted using the loading buffer containing glycerol once the volume of the sample is determined as glycerol makes the samples sink easily due to the density into the gel wells. The loading buffer also constitutes a tracking dye (bromophenol blue) which enables the researcher to track the position of samples inside the gel. To denature the proteins and break their secondary, tertiary and quaternary structures without disturbing the peptide bonds, the sample are heated after dilution. This denaturation helps in sustaining the transfer of negatively charged proteins under the influence of electric field. A positive and negative control is also needed for the sample and the purified protein sample or control lysates are used as positive controls [2, 4], which help in confirming the protein identity and the activity of antibody. β-actin/GAPDH is used as a negative control to confirm the nonspecific staining in the blots.

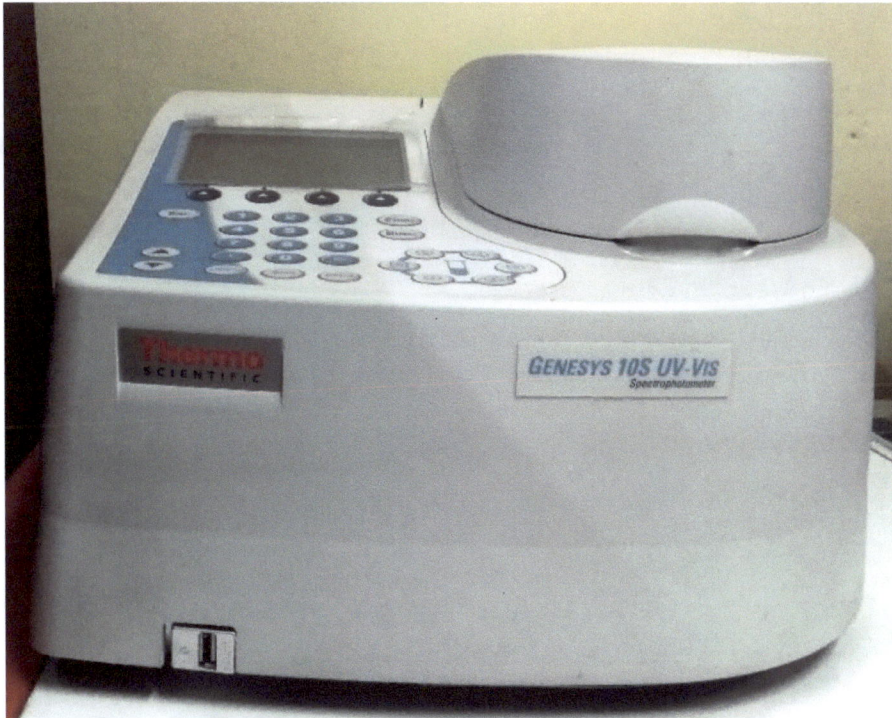

Fig. (1). Representation of Spectrophotometer.

GEL ELECTROPHORESIS

Two different types of polyacrylamide gels are used in WB *i.e.*, stacking and separating gel. The stacking gel is quite acidic (pH 6.8) and has lesser concentration of acrylamide, making the gel porous so the protein separates very poorly [5], while the resolving gel is comparatively basic (pH 8.8), with higher polyacrylamide content with narrow pore size. Therefore, the protein of interest is separated based on its size separating the smaller proteins faster than the larger proteins. When a voltage is applied, the loaded proteins having a negative charge on the gel will travel move toward the positive electrode. In general, the gels are prepared by using the solution containing Tris-HCl, SDS, Acrylamide/Bis-acrylamide, ammonium persulfate and Tetramethylethylenediamine (TEMED) as shown in Fig. (**2**). Along with samples and a marker loading into the wells, a sample buffer is also loaded. After the sample loading, the power supply is provided and the gel is run at a voltage that cannot overheat and distort the bands [4].

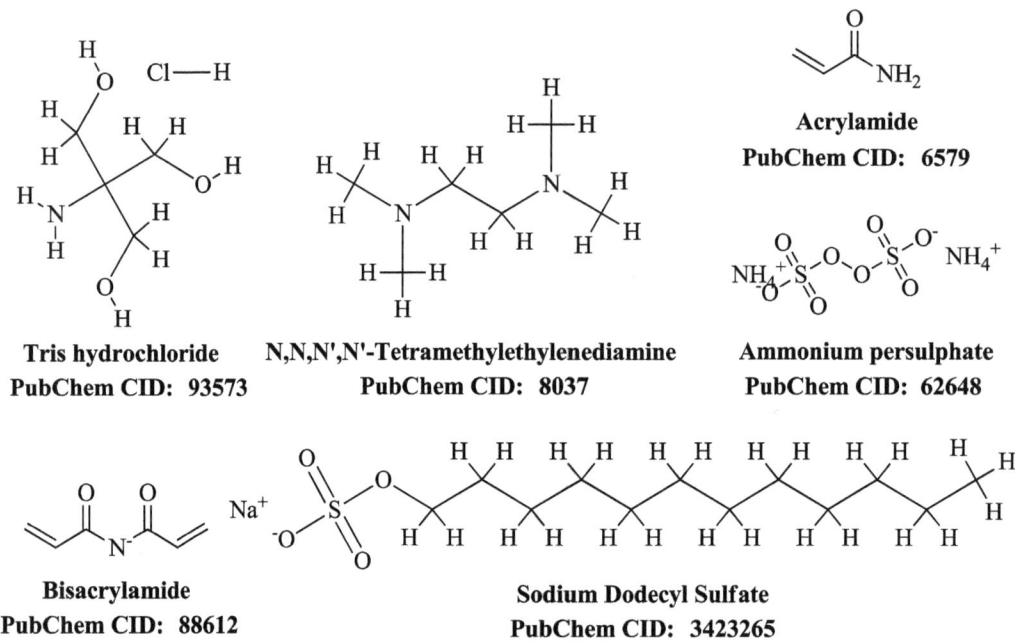

Acrylamide
PubChem CID: 6579

Tris hydrochloride
PubChem CID: 93573

N,N,N',N'-Tetramethylethylenediamine
PubChem CID: 8037

Ammonium persulphate
PubChem CID: 62648

Bisacrylamide
PubChem CID: 88612

Sodium Dodecyl Sulfate
PubChem CID: 3423265

Fig. (2). Chemical structure and Pubchem CID of Tris-HCl, Tetramethylethylenediamine, Acrylamide, Ammonium persulfate, Bis-acrylamide and SDS used in SDS-PAGE.

BLOTTING

After separating the protein on the resolving gel, it is transferred to a membrane. The transfer occurs perpendicular to the surface of the gel using an electric field (Fig. **3**), causing the proteins to move onto the membrane from the resolving gel [6]. The membrane is sandwiched between the resolving gel surface and the positive electrode, which includes a sponge pad at both ends, and filter papers in order to protect the blotting membrane and the gel. In order to have a clear image, a close contact between gel and membrane is vital besides the position of the membrane in between the resolving gel and the positive electrode (Fig. **3**). The membrane are kept in a way so the transfer of the negatively charged proteins occurs freely from the resolving gel on to the membrane and this type of transfer is known as electrophoretic transfer, which can be done in semi-dry or wet conditions. The wet conditions are preferred for larger proteins and are more reliable than semi-dry, as it is less likely to dry out the resolving gel. The membrane used should give solid support and plays essential role in the electrophoretic transfer process. Two types of membranes are generally used in WB *i.e.*, nitrocellulose and polyvinylidene difluoride (PVDF). Nitrocellulose membrane offers high affinity and retains the proteins but is quite brittle and doesn't offer re-probing. In this respect, PVDF membranes gain an upper hand due to its better mechanical support and for its re-probing and storage ability. It has got higher background possibility, which can be overcome by washing carefully.

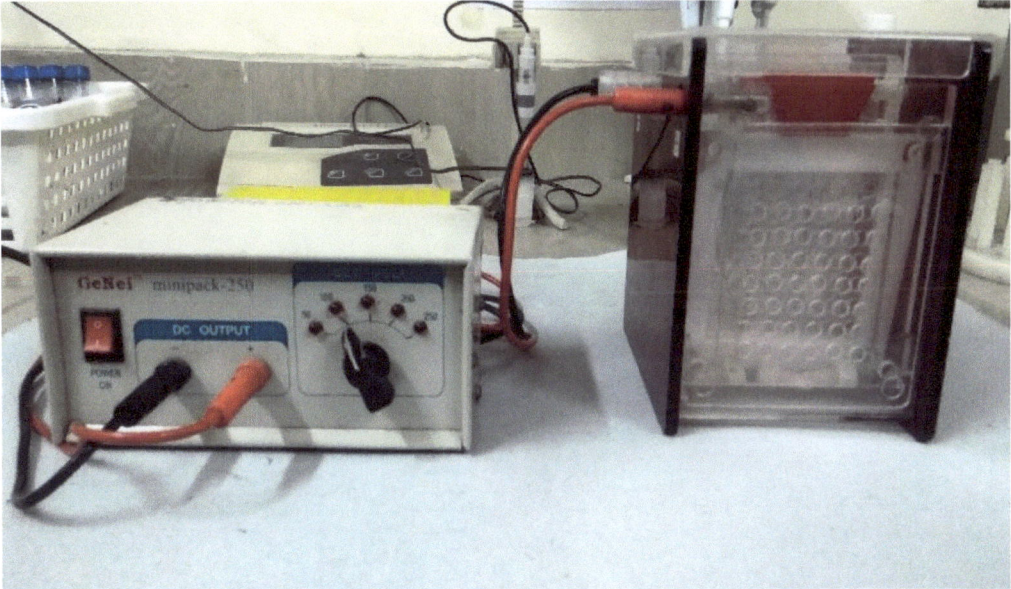

Fig. (3). Representation of Western blot unit.

WASHING, BLOCKING AND ANTIBODY INCUBATION

During western blotting, blocking is a very important step which helps in preventing the non-specific binding of antibodies to the membrane, which can be made possible with 5% BSA or non-fat dried milk diluted in TBST that reduces the background. Although non-fat dried milk is often preferred for its wide availability and inexpensive nature. Blocking solution should be used with utmost care as the milk proteins are not suitable for all detection labels. Thus, when the anti-phosphoprotein antibodies are used, the membrane is blocked using BSA because the casein protein present in milk can interfere with the results of WB [3]. Incubation of primary antibody with BSA seems to be a good strategy because of abundant need than the secondary antibody beside their reuse for next experiments. The antibody is diluted as per manufacturer instruction in a wash buffer, such as PBS or TBST and they are incubated along with the blots. These incubated membranes are washed accordingly to minimize background and to remove unbound antibody. These membranes are then treated with secondary antibody tagged with horseradish peroxidase (HRP) with the corresponding signal against specific protein of interest with signal captured on a X-ray film or by gel documentation system (Fig. **4**).

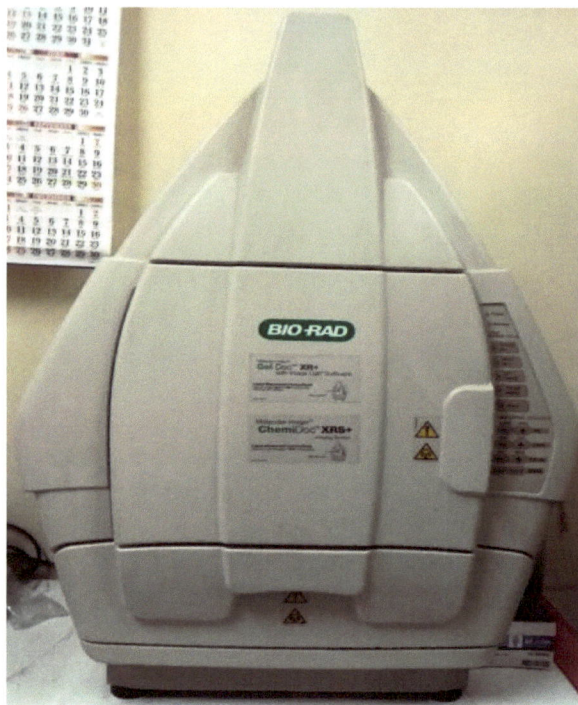

Fig. (4). Representation of Gel-Documentation system.

QUANTIFICATION

The quantification of the WB is usually said to be semi-quantitative (Fig. **5**) as it only tells about the relative comparisons of protein levels but the absolute quantity of the protein cannot be determined as there might be the variations in the loading and transfer rates of the samples in different lanes that are different on separate blots [3, 7]. Secondly, the signals generated are not linear across a range of concentrations in the samples used. Hence it should be used to model the concentration.

TROUBLESHOOTING

Although the process of western blotting is simple, many issues may arise leading to unexpected results. In general, these problems can be divided into five different sets, such as the absence of bands, pale bands or weak signal, unusual or unexpected bands, high background, and uneven spots on the blot. The absence of bands may occur due to several reasons ranging from antibody, antigen, to buffers used. The use of an improper antibody (primary or secondary) or their low concentration may result in the absence of a band on the blot beside it may also occur due to lower protein (antigen) concentration that can be confirmed through same protein from another source which confirms the real problem with the protein or an antibody.

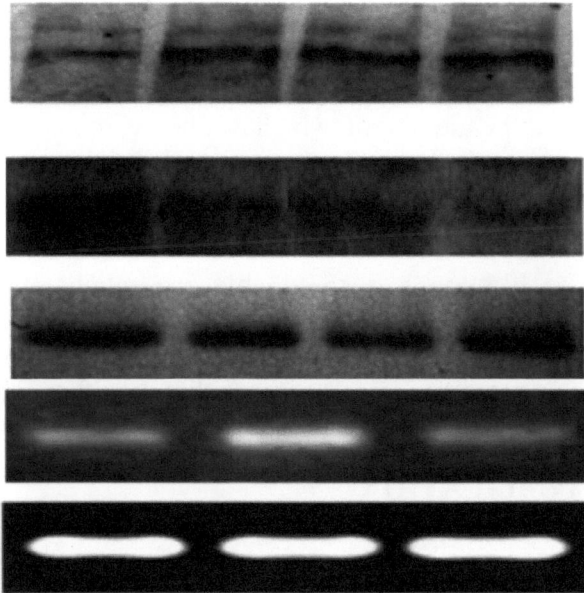

Fig. (5). Representation of some of the western blot images.

Pale bands occur due to multiple reasons such as high voltage or air bubbles during the transfer or multiple reuses of running buffer. These issues can be overcome by simple changes such as the use of low voltage or careful placement of membrane between the resolving gel surface and the positive electrode or by preparing fresh running buffer. The unusual or unexpected bands mainly occur due to compromised protein integrity by the proteases, for this, fresh sample should be used on ice in order to overcome this issue. When the position of the bands is high enough, their respective protein samples should be reheated in order to break the quaternary structure of the proteins. The non-flat bands, which occur due to the fast travelling of protein through the gel due to the low resistance, can be overcome by optimizing the gel used as per our sample requirements.

Additionally, prolonged washing with buffers may also lead to decreased signal beside these buffers being high reused and contaminated with sodium azide (inactivate HRP). By using new and non-contaminated buffers running, transfer, ECL and TBST, this problem can be overcome. The issues with the weak signals can also be overcome by enhancing the concentration of antibody or antigen or increasing their respective exposure time which may help in the visualization of clear bands on the blot. Weak signals may be caused due to the masking of antigens by the non-fat dry milk, in such a case it should be replaced with BSA or it has to be used in decreased amounts. Higher concentration of antibody and use of old buffers can also lead to the high background, which can be overcome by increased washing time (optimised). Sometimes, the uneven and patchy blots appear because of the improper transfer of the proteins or the air bubbles trapped in between the gel and the membrane which can be overcome by properly placing the membrane in between the gel and the positive electrodes.

CONCLUSION

In nutshell, Western blotting is a reasonable and reliable technique used for detection and quantification of protein. This chapter has covered the protocol, theory behind them, besides much-needed troubleshooting techniques, which may help young researchers to overcome the issues that they face during their wet lab practices of western blotting.

CONSENT FOR PUBLICATION

Not applicable.

CONFLICT OF INTEREST

The author confirms that this chapter contents have no conflict of interest.

ACKNOWLEDGEMENTS

Declare none.

REFERENCES

[1]   Konradi CL. Quantification of protein in brain tissue by western immunoblot analysis In Drugs of Abuse. Humana Press 2003; pp. 263-71.

[2]   Feist P, Hummon AB. Proteomic challenges: sample preparation techniques for microgram-quantity protein analysis from biological samples. Int J Mol Sci 2015; 16(2): 3537-63.
[http://dx.doi.org/10.3390/ijms16023537] [PMID: 25664860]

[3]   Mahmood T, Yang PC. Western blot: technique, theory, and trouble shooting. N Am J Med Sci 2012; 4(9): 429-34.
[http://dx.doi.org/10.4103/1947-2714.100998] [PMID: 23050259]

[4]   Iqbal J, Ahmad R. Western blot: proteins separating technique, protocol, theory and trouble shooting. J Bacteriol Mycol Open Access 2017; 4(2): 62-7.
[http://dx.doi.org/10.15406/jbmoa.2017.04.00088]

[5]   Rao VS. Transgenic Herbicide Resistance in Plants. 1st Edition.. Crc Press 2014; p. 486 pages. ISBN: 9780429168925.

[http://dx.doi.org/10.1201/b17857]

[6]   Jin S. Development of Microfluidic Based Western Blot Technology for Fast and High-Content Analyses 2015.http://hdl.handle.net/2027.42/116663

[7]   Taylor SC, Posch A. The design of a quantitative western blot experiment. BioMed Res Inter 2014; 2014
[http://dx.doi.org/10.1155/2014/361590]

CHAPTER 5

Immunohistochemistry as an Important Technique in Experimental and Clinical Practices

Hareram Birla[1], Sachchida Nand Rai[1], Saumitra Sen Singh[1], Walia Zahra[1], Neeraj Tiwari[1], Aijaz A. Naik[2], Anamika Misra[3], Shikha Bharati[4] and Surya Pratap Singh[1],*

[1] *Department of Biochemistry, Institute of Science, Banaras Hindu University, Varanasi-221005, India*

[2] *School of Studies in Neuroscience, Jiwaji University, Gwalior-474011, India*

[3] *Institute of Medical Science, Banaras Hindu University, Varanasi-221005,India*

[4] *School of Life Science, Jawaharlal Nehru University, New Delhi, 110067, India*

Abstract: Immunohistochemistry (IHC) is a well-known technique in the field of biological and medical sciences. This technique is based on the principle of antigen-antibody interaction and is used for identification of cellular or tissue constituents, *i.e.*, an antigen by using a specific antibody. The binding of an antibody to an antigen is confirmed either by labelled primary antibody itself or by using secondary labelling method such as fluorescence labelled antibody. Such interactions give information about the cellular process occurring inside the cell. In last few years, huge amount of data have been generated using IHC. Furthermore, adequate knowledge of this technique is required for the optimum result and its reproducibility. The detailed information about the tissue section, antigen retrieval (AR), increased sensitivity of the detection systems and proper standardization are the key points for this technique. This protocol will address overview of the technique, tissue preparation, microtome, antigen retrieval, antibodies and antigen fixation, detection methods, background reduction and trouble shootings.

Keywords: Antibody, Antigen, DAB, Diagnosis, Enzyme, Fixation, Formaldehyde, Fluorescence, GFAP, Histology, IHC, Immunofluorescence, Microtome, Pathology, Tissue.

INTRODUCTION

Immunohistochemistry (IHC) is a well-acknowledged technique in the field of biological and medical sciences [1]. This technique is widely used for diagnosis of

---

Correspondence: **Corresponding Author Surya Pratap Singh**: Department of Biochemistry, Institute of Science, Banaras Hindu University, Varanasi, India; E-mail: suryasinghbhu16@gmail.com

various diseases and detection of cellular antigen (Ag) in various tissue samples [2]. IHC technique is based on three scientific disciplines: immunology, histology and chemistry. This technique works on the principle of antigen-antibody (Ag-Ab) interaction and is used for the identification of cellular or tissue constituents, *i.e.*, an Ag by using a specific antibody (Ab). The binding of Ag to an Ab is confirmed either by labelled antibody itself or by using secondary labelling methods such as fluorescent labelled secondary Ab. Furthermore, the Ag-Ab interaction can also be recognized by coloured histochemical reaction under light microscope and fluorochromes using ultraviolet light [3, 4]. Also, proper standardization and inclusion of negative control are essential for the reproducibility of results. Pathologists and researchers encounter various challenges due to diversity of tissues and cross-reactivity of antibody among different species. This technique is highly reliable and accepted for routine diagnostic and research activities. The interaction between Ag and Ab confirms the presence of epitope at cellular level. The technique has changed the scenario of experimental and clinical pathology. One excellent example is tumour heterogeneity in which epithelial-to-mesenchymal transition has been reported using IHC. The alteration of different proteins can be determined by expression microarrays. However, the differential expression of specific Ag in glandular epithelial and spindle cell populations can be elucidated specifically through IHC. Presently, the IHC is utilized not only for diagnostic purposes but also for identification of some prognostic markers. Such markers help in the selection of specific treatment [4].

DIRECT TECHNIQUE

In traditional direct IHC, labelled primary antibody is used. For labelling either fluorochrome (more common) or an enzyme is used as a label. In this method, a direct reaction occurs between the labelled antibody and antigen present in the histological or cytological preparation. Thus this technique is very quick and can be used easily. However, on the other hand this technique shows low accuracy, sensitivity and signal amplification [5].

NEW DIRECT TECHNIQUE (ENHANCED POLYMER ONE STEPS TRAINING METHOD)

Commercially, this method is known as 'Enhanced Polymer One-Step Staining (EPOS)', from Dakocytomation [6]. In this method, signal amplification and sensitivity are enhanced by attaching a large number of Ab molecules and peroxidase enzyme to a dextran polymer backbone. However, this method is not frequently used perhaps due to a limited number of primary Abs available [7].

TWO-STEP INDIRECT TECHNIQUE

This method is highly sensitive as compared to the traditional direct technique. Here, an un-labelled primary Ab is visualized by a labelled secondary Ab against the immunoglobulin of the animal species in which the primary antibody has been raised. The enzyme Horseradish peroxidase together with an appropriate chromogen substrate is usually used for the labelling. Thus this technique offers diversity in terms of labelled secondary Ab and can be used with a variety of primary Ab raised in the same animal species [7].

POLYMER CHAIN TWO STEP INDIRECT TECHNIQUE

In this technique, an unconjugated Ab is used against a particular antigen. Thereafter, secondary Ab conjugated with an enzyme such as Horseradish peroxidase labelled to polymer such as forex, dextran chain is used. Conjugation of secondary Ab raised in two different animals species, such as anti-mouse and anti-rabbit allows researchers to use the same reagent for both monoclonal (rabbit and mouse) and polyclonal (rabbit) primary antibody [7].

UNLABELED ANTIBODY ENZYME COMPLEX TECHNIQUES (PAP AND APAP)

Nowadays, a soluble peroxidase anti-peroxidase complex (PAP) is used frequently as compared to conventional immunoenzyme bridge method [7] shown in Fig. (**1**).

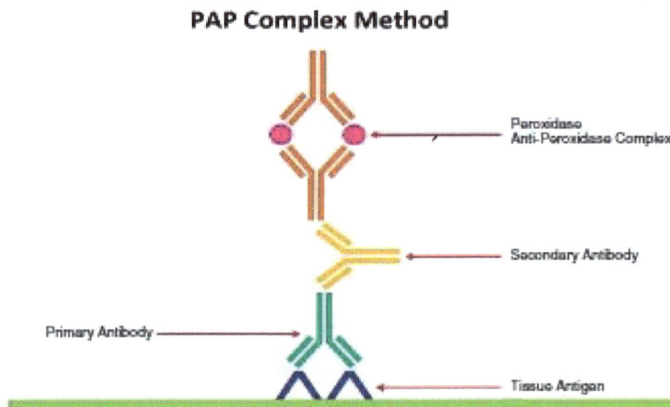

PAP Complex Method

Peroxidase Anti-Peroxidase Complex

Secondary Antibody

Primary Antibody

Tissue Antigen

Fig. (1). Schematic representation of the PAP complex method.

For this tissue section is initially treated with primary antiserum and then Ag is identified by sequential application of purified specific rabbit antiserum, a soluble Horseradish peroxidase-anti Horseradish peroxidase (PAP), 3,3'diaminobenzidine, hydrogen peroxide and osmium peroxide.

High yield of PAP is prepared by precipitating rabbit antiserum with Horseradish peroxidase at equivalence and then washed precipitate is solubilized using excess P0 at pH 2.3, 1°C followed by instant neutralization and separation of PAP from P0 by half saturation using ammonium sulfate [8]. Similarly, alkaline phosphatase- anti-alkaline phosphatase complexes (APAP) can be formed through the same principle using mouse to raise alkaline phosphatase Ab. Cordell has first developed this method in 1984 [7].

IMMUNOGOLD SILVER STAINING TECHNIQUE (IGSS)

Faulk and Taylor first introduced colloidal gold as a label for IHC in 1971. It has been widely used in ultra-structural immuno-localization and can be used in both direct and indirect methods. In this technique, metallic silver layer is added to enhance gold particles by producing a metallic silver precipitate that overlays the colloidal gold marker. This can be observed under a light microscope. In a protective colloid gum of Arabic or Silver, lactate is used as an ion supplier and hydroquinone as a reducing agent at pH 3.5. This method is more sensitive than PAP. However, formation of silver deposits in the background is a drawback, which can create confusion when small amount of Ag has to be identified [9]. This method is more economically feasible than immune-peroxidase method as it uses more diluted primary antisera. Substantial volumes of reagent can be prepared easily by a technician in just a few hours. Minimal cost is required for preparing these reagents apart from antisera. The other aspect of economic feasibility is that much higher diluted secondary antisera adsorbed to colloidal gold are used as compared to immuno-peroxidase methods [4]. Immunogold-silver staining is represented schematically in the following Fig. (2).

For this, tissue sections are initially treated with primary antiserum raised in rabbits directed against the human immunoglobulin (Ig) under study. Colloidal gold adsorbed to antiserum directed against rabbit immunoglobulins (Igs) is added. Gold particles introduced to antigenic sites are subsequently revealed by the silver precipitation reaction [4].

SILVER
PRECIPITATE

GOLD PARTICLE

SWINE ANTI-RABBIT Igs

RABBIT ANTI-HUMAN Ig

HUMAN Ig

TISSUE SECTION

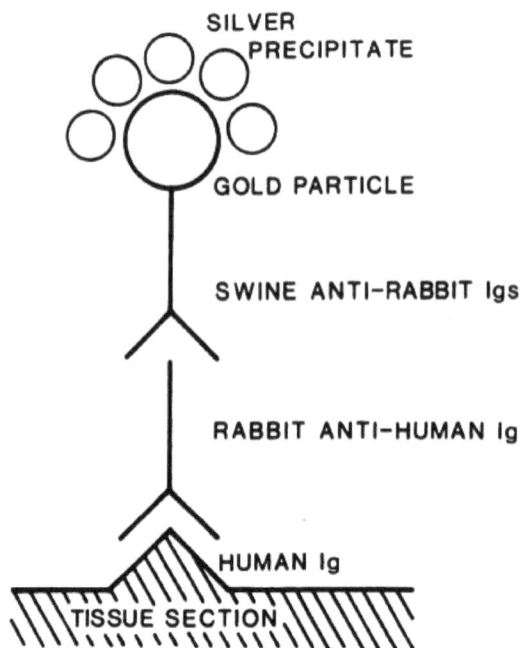

Fig. (2). Schematic representation of the IGSS method.

THE PRINCIPLE OF IHC

Though, the IHC technique existed since 1930. However, it has been first reported in 1941 by Coons and his colleagues for identification of pneumococcal Ag using Fluorescein isothiocyanate (FITC)-labeled antibody with a fluorescent dye. The technique basically aims for the identification of various biological markers using a small amount of Ab without causing any damage to the cell or tissue. With the advancement, various enzyme labels were introduced for identification of Ag such as alkaline phosphatase and peroxidase [7, 8]. The colloidal gold label has been also widely used for observation of immunohistochemical reactions in light and electron microscope. Radioactive elements can also be used to see immunoreactions by the means of autoradiography [9].

TISSUE PREPARATION

Tissue and cellular sample for IHC/ICC study are prepared freshly and appropriately. After appropriate incubation steps, tissue is coronally cut into thin sections (5-15 μm). Also, sample preparation is closely linked to tissue fixation and its identification by using suitable detection techniques (fluorescence versus chromogen). Therefore, precautions should be taken at each step. Small block of tissue not more than 2 cm square and 5mm thick should be prepared. Mostly, tissue preparation is determined by one experiment variable. For example, when tissues are immersion-fixed using paraformaldehyde and further embedded in paraffin. The section cutting of such samples are done in microtome. On the other hand, if phosphorylation-dependent epitope is to be detected the tissue should be frozen instantly and sectioning should be done in cryostat as shown in Fig. (**3**) [10].

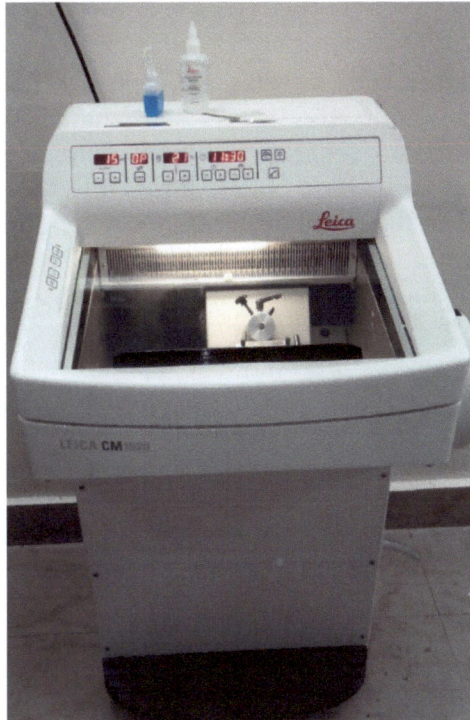

Fig. (3). Leica CM1520 Cryostat machine for Standard Applications in the Clinical Histopathology Laboratory.

The IHC technique includes the following steps:-

I. Deparaffinization of tissue sections akin on polylysine-coated slides (or else the aqueous solutions do not penetrate).
II. Quenching of endogenous enzymes (which otherwise react with IHC reagents give false positive results). Alkaline phosphatases, peroxidases and biotin are examples of these endogenous enzymes. This is usually blocked by 3% H_2O_2 or with free avidin.
III. Antigen retrieval.
IV. Blocking of nonspecific binding sites.
V. Binding of primary antibody.

VI. Binding with the biotinylated secondary antibody.
VII. Detection methods using peroxidases-anti-peroxidase methods, avidin-biotin conjugates, peroxidases complexes or the more recently polymer labelling two-steps method.
VIII. Addition of chromogen substrate usually DAB.
IX. Counterstaining, dehydrating and cover slipping the slide [6].

ANTIGEN RETRIEVAL

Different pre-treatment methods of the sample can affect the Ag differently. Confirmation of protein molecules are altered during traditional procedures involving fixation and paraffin-embedding. Due to this, the Ag-Ab interaction can be negatively affected and decreases the intensity of IHC reaction. Though, the confirmation of Ab used in the procedure is preserved in its native form. Also, the confirmation of Ag located in the tissue might be affected differently. The Ag may encounter modification in conformation during aldehyde fixation [8]. Therefore, Ag retrieval becomes necessary in order to recover the antigenicity of tissue sections masked by formalin fixation and paraffin embedding [11]. Several techniques like microwave oven radiation, proteolytic enzyme digestion, pressure cooker heating, autoclave heating, water bath heating and steamer heating can help in revealing the antigenicity of tissue sections [7].

A. Proteolytic Enzyme Digestion

A solution of enzymes such as protease, trypsin and pepsin are used for epitope unmasking through enzymatic digestion of tissue sections. Though, most of the Ag retrieval methods of are strongly effective and might result in elevated background. Generally, less intense staining is provided by enzyme digestion as compared to heating [12]. Huang first of all introduced the method of protease-induced epitope retrieval (PIER). For this method, many enzymes such as proteinase K, trypsin, pepsin, proteinase and ficin have been used [18]. PIER perhaps involves the digestion of protein. This action is non-specific and might negatively affect the Ag. The duration of fixation, incubation parameters (time, pH and temperature), and the concentration of enzymes are important for PIER. There is an inverse relationship between the duration of digestion and the fixation time [8]. Proteolytic digestion sometimes can cause a detrimental effect on the antigenicity of some Ag, which might produce false negative and false positive results. Thus, an adequate balance between under and over-digestion is important during the proteolytic enzyme digestion [7].

B. Heat Mediated Antigen Retrieval Techniques

Ag fixed in cross-linking fixatives, such as formaldehyde is immunohis- to chemically detected by a method known as the heat-induced epitope retrieval (HIER). The basis of HIER action is unknown. However, it helps in reverting the conformation of proteins affected during fixation. Epitope can be also be unmasked through heating by breaking the methylene crosslinks. Furthermore, there is possibility of another mechanism also. The immunostaining of tissue fixed in ethanol can be enhanced using this method. Extraction of diffusible blocking proteins, precipitation of proteins, rehydration of the tissue section allows better penetration of Ab. The heat mobilization of trace paraffin is other hypotheses proposed in this method [8]. Morgan *et al.*, (1997) have postulated another possible theory in which Ab is prevented by calcium coordination complex formed during formalin fixation. Calcium coordination complexes are formed through interaction of hydroxyl-methyl group and other unreacted oxygen-rich groups, such as carboxyl or phosphoryl with calcium. Some of the calcium coordinate bonds might get weakened by providing high temperature [1]. Different equipment such as vegetable steamer, pressure cooker and microwave oven or a de-cloaker (commercial pressure cooker with electronic controls for temperature and time) are used frequently [1]. The de-cloaker has the advantage over other heating devices because it is not affected by the atmospheric pressure.

ANTIGEN-ANTIBODY INTERACTION

The IHC technique is based on Ag-Ab interaction and is used for the identification of the cellular and tissue-components. The bound Ab with Ag is either identified directly by a labelled primary Ab or by using a secondary labelling method [1]. The staining involved in IHC uses fluorophore-labelled (immunofluorescence) and enzyme-labelled (immune-peroxidase) Ab for identification of different proteins and other cellular constituents. Immuno-peroxidase method is widely used in surgical pathology which cannot be retrieved by H&E staining [13]. In last few decades, different Ab, various tools and detection systems have been developed. These efforts are further used for the detection of various protein markers [14]. The increasing demand of surgical pathology increased the discovery of new biomolecules that can be used diagnosis and targeted therapy [15]. The Ab for immuno-histological testing is chosen on the basis of tumour specificity and tumour under evaluation. The positive Ag-Ab reaction is identified by several detection systems. For increasing sensitivity labelled secondary antibody is used against the primary antibody. Relatively large number of enzyme molecules, *e.g.* peroxidase can be attached to the Ag for increasing the sensitive. On the basis of enzymatic reaction, the selection of different precipitating chromogen are selected such as diaminobenzidine (brown) or aminoethyl-carbazole (red) [12].

AVIDIN–BIOTIN COMPLEX

Various Avidin-Biotin Complex (ABC) techniques have been evolved during recent years. The high affinity of streptavidin (from *Streptomyces avidinii*) and avidin (from the chicken egg) for biotin forms the basic principle of ABC method. Biotin is a natural vitamin (vitamin B7, vitamin H). Generally, four moles of biotin binds to one mole of avidin. The strongest known non-covalent interaction is the avidin-biotin interaction. The formation of bond between avidin and biotin is highly rapid and is unaffected by variation in pH. Nowadays, avidin has been replaced by streptavidin, as it has a high affinity for binding with lectin-like and other negatively charged tissue-components. ABC method can be effectively used in light, fluorescence and electron microscopy. ABC reagents are generally added in the following sequence: (1) primary antibody is directed toward a specific determinant on the cells, (2) biotinylated secondary antibody, and (3) labelled streptavidin. "Preformed complex" method (vectors laboratory) is the recent and most widely used ABC technique. A preformed complex of streptavidin and biotinylated enzyme is added after the application of biotinylated primary or secondary Ab. This latest technique is considered most sensitive in many ABC applications. However, the Ag is present in lower amounts or when the cost of primary Ab is significantly high [5].

HAPTEN LABELLING TECHNIQUE

Haptens, such as dinitrophenol and arsanilic acid are highly used for bridging in IHC. The hapten is usually linked to a primary Ab and a complex is formed with an anti-hapten. This can be either hapten labelled enzyme or hapten-labelled PAP complex [16].

DETECTION METHODS

Different manufacturers have offered a number of good quality detection systems. However, the peroxidase-based detection systems are about four times more sensitive than alkaline phosphatase. The replacement of older avidin-based materials with newer streptavidin or neutravidin reagents has relatively improved the results with significant reduction of backgrounds. Furthermore, the sensitivity of detection system also varies from one reagent to others. Reagents are provided by the manufacturers either in "ready to use" format or in concentrated form which is comparatively more economical. It is critical to use a peroxidase-stabilizing diluent to dilute the concentrated detection complex to its working dilution. Depending upon the preference, a number of different chromogens can be used. We generally prefer Diaminobenzidine (DAB), which gives a brown reaction product insoluble in organic solvents (Fig. **4**). It can be easily mounted in standard mounting media without any decrease in signal over time. Aminoethylcarbizole (AEC), is another widely used reagent which gives red signal. But, in our laboratory, we usually don't prefer it for a number of reasons. During experiments, we have observed that AEC is less sensitive as compared to DAB. This is because AEC requires an aqueous mounting media. Further, upon high power examination this gives a fuzzy appearance. Also, on prolonged slide storage, AEC has the tendency to get deteriorated easily. We prefer DAB in such cases. DAB is also preferred for the interpretation of individual cell in a mixed cell population. We have also evaluated another chromogen that can be mounted in standard mounting media (vector red) with red-brown reaction product. Chromogen development steps can be performed using several options. However, placing the slides on flat transparent sheets and flooding them with an abundant amount of Ag is very common. Chromogen development can be observed under the light microscope. After optimum time the reaction can be stopped by placing the slides under running tap water. However, it takes too much time and thus DAB staining is much preferred.

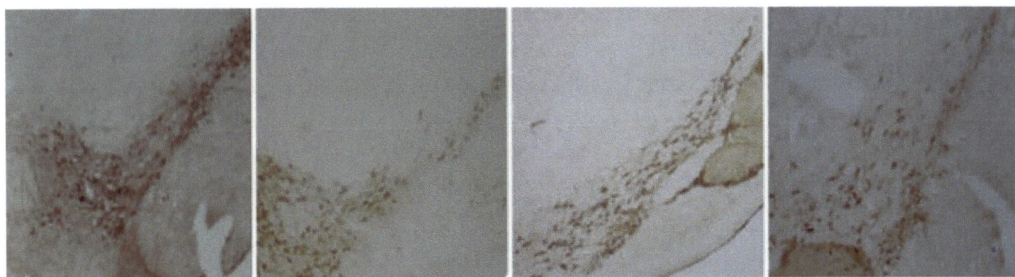

Fig. (4). TH immunoreactivity of dopaminergic neurons in mouse substantianigra [17].

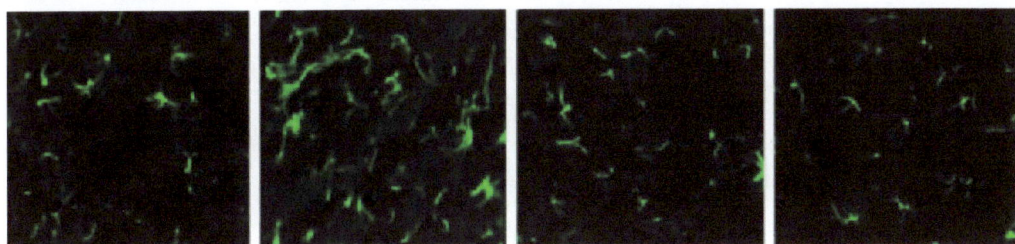

Fig. (5). Immunofluorescence staining GFAP (Glial fibrillary acidic protein) in striatum [18].

Interestingly, it has been observed that the results through DAB staining can be substantially improved by pre-warming the DAB in an oven. In our laboratory, the optimal temperature has been observed between 23-28°C for better results. Furthermore, an increased background can be observed if the DAB is too hot (around 35-40 °C).

QUANTIFICATION OF THE IMMUNOHISTOCHEMICAL STAINING

The quantification of the IHC is done either in terms of total number of immunopositive cells or the intensity of staining signal. Such analysis can be performed by softwares like ImageJ (a freely available software from National Institute of Health (NIH)) or commercial quantification application software's available for different microscopes [19, 20]. The quantification data can be also subjected to statistical analysis for evaluating the differences in expression of antigen in control and experimental group. Furthermore, for the quantification images should be of the same magnification and utmost care should be taken to keep the same set of exposure and saturation for homogeneity and consistency.

APPLICATION AND IMPORTANCE

IHC uses specific Ag-Ab interactions and is more beneficial than the conventional enzyme based techniques as these enzymes can only identify a limited number of proteins. Therefore, IHC has extended attention in the fields of medical diagnostics and for various research activities [21]. The applications include: identification of morphologically non-differentiated neoplasias (*E.g.*: lymphoma), primary site of malignant neoplasias, prognostic factors and therapeutic indications of several diseases, discrimination between the malignant versus benign tumour on the basis of cell proliferation [11 - 13].

IHC is highly effective as compared to H&E diagnosis in majority of tumour cases. IHC is typically applied to such cases when the definitive diagnosis cannot be established alone on the basis of H&E staining [19]. In a routine pathology, the IHC is utilized for differentiation of carcinoma, lymphoma and melanoma [6]. IHC staining is also very useful in the identification of genetic mutations, cancers of unknown origin, chromosomal translocations, neurodegenerative disease markers identification, tubular and lobular carcinomas of breast cancer (Fig. **6**).

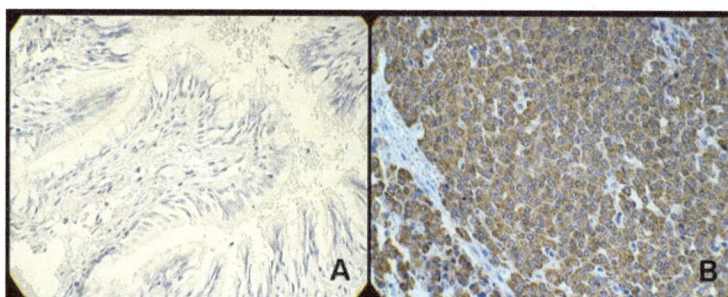

Fig. (6). Heparanase expression in the diagnosis of bronchopulmonarycarcinoid tumours. Optical microscopy at X400 power: A) negative expression of heparanase (absence of staining—peroxidase—in cell's cytoplasm) in bronchial mucosa not compromised by neoplasm; B) positive expression of heparanase (presence of cytoplasm full of peroxidase—brownish areas) in bronchopulmonarycarcinoid tumour [22].

TROUBLES AND LIMITATIONS

IHC is a simple technique and its outcome depends on different factors. Primarily, the success of IHC depends on the experience and observation [20]. Therefore, this technique requires rigorous knowledge of cases and results should be interpreted cautiously. Precautions should be taken at each step such as specimen fixation, tissue preparation, Ab selection, detection and result interpretation. Protocol can be modified at different step in order to minimize the undesirable effects. Furthermore, the selection of primary, secondary and tertiary anti-sera are important. Also, the dilutions of anti-sera must be standardized for optimal results.

TROUBLESHOOTING OF THE EXPERIMENTAL PROBLEMS

The failure of staining possibly suggests that the primary Ab was not added at all or the secondary Ab used was from the wrong species. Also, the addition of reagents was not done in the proper order. The sections got dried while performing the experiment or the labelled reagent was not added at all. Another problem might be related to the staining of positive controls and no staining in test section. This might be due to the Ag present in the section has masked or destroyed during the procedure. This can be overcome by increasing the incubation time with primary Ab. Other problems may include the fixation of tissues harsh fixative or subjected to high temperature. Sometimes, excessive background staining is seen in experimental as well as control slides. This might be due to following reasons: primary Ab concentration was high, the sections were not properly deparaffinised, incubation with primary or secondary Ab was for the longer time period, or the washing of the slides was not proper. To overcome this problem, the blocking reagent for endogenous peroxidase should be used properly. Moreover, if background staining was observed only in the unknown sample, it can be possible that the first section has undergone excessive necrosis or autolysis.

IMMUNOENZYMATIC LABELLING PROTOCOL FOR LOCALIZA-TION OF ASTROCYTES USING ANTI-GFAP ANTIBODY FOR CRYOSTAT SECTIONS DAY 1 PROCESSING

Air dry the cryocut brain tissue sections at 37°C for 1 hour.

1. Wash the sections in phosphate buffer saline (1x), 3 changes of 5 min each.
2. Treat the sections with 0.1% Triton X-100 in PBS for 10 min for membrane permeabilization and antigen retrieval.
3. Wash the sections in PBS, 3 changes of 5 min each.

4. **Endogenous blocking**: For endogenous blocking use 1% Hydrogen peroxide (H_2O_2) in PBS for 30 min.
5. Wash the sections in PBS, 3 changes of 5 min each.
6. **Non-specific blocking**: For non-specific protein blocking, incubate the sections with 1% Serum (Use Rb. specific Vector kit PK6101 or other available blocking serum or Kits) prepared in PBS for 90 min at room temperature.
7. **Primary Antibody Incubation:** Incubate the sections with primary antibody **(anti-GFAP, Rabbit polyclonal, Wako or available anti-GFAP antisera)** at **dilution 1:1000** (diluted in 1% BSA in 1x PBS) for overnight at 4°C in a humid chamber.

DAY 2 PROCESSING

1. Bring sections to room temperature.
2. Rinse the sections in 1x PBS, 3 changes of 5 min each to remove excessive unbound primary anti-GFAP antibody.
3. **Secondary Ab incubation**:
   Incubate the sections with secondary Ab(from Vector Kit PK6101 or available choice secondary antisera) at a dilution of 1: 100 (diluted in 1% BSA in PBS) at room temperature for 90 min.
4. Rinse the sections in PBS, 3 changes of 5 min each.
5. **Tertiary Ab incubation:** Incubate the sections with tertiary Ab. {A+B reagents from the vector kit PK6101} at a dilution of 1: 200 (diluted in 1% BSA in PBS + 0.1% Tween 20) at room temperature for 90 min.
6. Rinse in PBS, 3 changes of 5 min each at room temperature to remove unbound secondary antibody.
7. Visualize with DAB solution (25 mg DAB in 100 ml of PBS + 60 ml H_2O_2 for 20 min at room temperature. (The time may vary and slides should be carefully observed and once the desired colour intensity is achieved, the reactions should be terminated). Avoid direct contact with DAB as it is carcinogenic.
8. Wash the slides in slow running tap water at room temperature for 15 min.
9. Rinse in distilled water (2 changes for 5 min each).
10. Air dry slides for 1 hour at 37°C.
11. Dehydrate slides in 100% Alcohol for 10 min.
12. Clear in Xylene for 10 min.
13. Mount in DPX and observe the slides as shown in Fig. (**5**).

IMMUNOHISTOCHEMICAL PROTOCOL FOR LABELLING ASTRO-CYTES USING ANTI-GFAP BY IMMUNE-FLUORESCENCE METHOD FOR CRYOSTAT SECTIONS DAY 1 PROCESSING

1. Air dry the cryocut sections for 1 hour at room temperature.

2. Wash the sections in phosphate buffer saline (PBS) 3 times for 5 min each.
3. Treat the sections with 0.5% Triton X-100 in PBS for 20 min.
4. Wash the sections in PBS three times for 5 min each.
5. **Protein blocking:** For non-specific protein blocking, incubate the sections using 10% normal goat serum (NGS) in 1x PBS for 90 min at room temperature.
6. **Apply Primary antisera**: Use anti-GFAP (Rabbit polyclonal, Dako) at a dilution of 1:1000 (diluted in 5% BSA in PBS + 0.5% Triton X 100) and incubate the sections at 4°C overnight in a humidified chamber.

DAY 2 PROCESSING

1. Bring the sections to room temperature
2. Wash the sections in PBS three times for five min each.
3. **Apply Secondary antisera:** Use anti-Rabbit FITC, Goat raised or another secondary anti-serum of your choice and availability, at a dilution of 1:300 (diluted in 5% BSA in PBS + 0.5% Triton X-100) for 2 hours at room temperature in a dark humid chamber. Avoid direct exposure of sections to bright light.
4. Wash the sections in PBS five times for ten min each to ensure proper washing of the unbound secondary anti-sera.
5. Mount in vector Hardset + DAPI mounting medium or available mounting media.
6. Observe the slides under a fluorescent microscope to see astrocytes, their distribution, phenotype and other associated profiles.

CONSENT FOR PUBLICATION

Not applicable.

CONFLICT OF INTEREST

The author confirms that this chapter contents have no conflict of interest.

ACKNOWLEDGEMENTS

Declared none.

REFERENCES

[1]   Shin D, Arthur G, Popescu M, Korkin D, Shyu CR. Uncovering influence links in molecular knowledge networks to streamline personalized medicine. J Biomed Inform 2014; 52: 394-405.
[http://dx.doi.org/10.1016/j.jbi.2014.08.003] [PMID: 25150201]

[2]   Fredolini C, Byström S, Pin E, *et al.* Immunocapture strategies in translational proteomics. Expert Rev Proteomics 2016; 13(1): 83-98.
[http://dx.doi.org/10.1586/14789450.2016.1111141] [PMID: 26558424]

[3]   Bordeaux J, Welsh A, Agarwal S, *et al.* Antibody validation. Biotechniques 2010; 48(3): 197-209.
       [http://dx.doi.org/10.2144/000113382] [PMID: 20359301]

[4]   Braun E, Eichen Y, Sivan U, Ben-Yoseph G. DNA-templated assembly and electrode attachment of a
       conducting silver wire. Nature 1998; 391(6669): 775-8.
       [http://dx.doi.org/10.1038/35826] [PMID: 9486645]

[5]   Daneshtalab N, Doré JJ, Smeda JS. Troubleshooting tissue specificity and antibody selection:
       Procedures in immunohistochemical studies. J Pharmacol Toxicol Methods 2010; 61(2): 127-35.
       [http://dx.doi.org/10.1016/j.vascn.2009.12.002] [PMID: 20035892]

[6]   Gerstein AS. Molecular biology problem solver: a laboratory guide. John Wiley & Sons 2004.

[7]   Bancroft JD, Gamble M, Eds. Theory and practice of histological techniques. Elsevier health sciences
       2008.

[8]   Ramos-Vara JA. Technical aspects of immunohistochemistry. Vet Pathol 2005; 42(4): 405-26.
       [http://dx.doi.org/10.1354/vp.42-4-405] [PMID: 16006601]

[9]   Mason DY, Sammons R. Alkaline phosphatase and peroxidase for double immunoenzymatic labelling
       of cellular constituents. J Clin Pathol 1978; 31(5): 454-60.
       [http://dx.doi.org/10.1136/jcp.31.5.454] [PMID: 77279]

[10]   Taylor CR, Levenson RM. Quantification of immunohistochemistry--issues concerning methods,
       utility and semiquantitative assessment II. Histopathology 2006; 49(4): 411-24.
       [http://dx.doi.org/10.1111/j.1365-2559.2006.02513.x] [PMID: 16978205]

[11]   Hayat MA. Microscopy, immunohistochemistry, and antigen retrieval methods: for light and electron
       microscopy. Springer Science & Business Media 2002.

[12]   Freitas TM, Miguel MC, Silveira ÉJ, Freitas RA, Galvão HC. Assessment of angiogenic markers in
       oral hemangiomas and pyogenic granulomas. Exp Mol Pathol 2005; 79(1): 79-85.
       [http://dx.doi.org/10.1016/j.yexmp.2005.02.006] [PMID: 16005715]

[13]   Idikio HA. Immunohistochemistry in diagnostic surgical pathology: contributions of protein life-cycle,
       use of evidence-based methods and data normalization on interpretation of immunohistochemical
       stains. Int J Clin Exp Pathol 2009; 3(2): 169-76.
       [PMID: 20126585]

[14]   Gown AM. Genogenic immunohistochemistry: a new era in diagnostic immunohistochemistry. Curr
       Diagn Pathol 2002; 8(3): 193-200.
       [http://dx.doi.org/10.1054/cdip.2002.0116]

[15]   Jambhekar NA, Chaturvedi AC, Madur BP. Immunohistochemistry in surgical pathology practice: a
       current perspective of a simple, powerful, yet complex, tool. Indian J Pathol Microbiol 2008; 51(1): 2-
       11.
       [http://dx.doi.org/10.4103/0377-4929.40382] [PMID: 18417841]

[16]   Jawhar NM. Tissue Microarray: A rapidly evolving diagnostic and research tool. Ann Saudi Med
       2009; 29(2): 123-7.
       [http://dx.doi.org/10.4103/0256-4947.51806] [PMID: 19318744]

[17]   Yadav SK, Prakash J, Chouhan S, *et al.* Comparison of the neuroprotective potential of Mucuna
       pruriens seed extract with estrogen in 1-methyl-4-phenyl-1,2,3,6-tetrahydropyridine (MPTP)-induced
       PD mice model. Neurochem Int 2014; 65: 1-13.
       [http://dx.doi.org/10.1016/j.neuint.2013.12.001] [PMID: 24333323]

[18]   Ojha S, Javed H, Azimullah S, Abul Khair SB, Haque ME. Neuroprotective potential of ferulic acid in
       the rotenone model of Parkinson's disease. Drug Des Devel Ther 2015; 9: 5499-510.
       [PMID: 26504373]

[19]   Jordan RC, Daniels TE, Greenspan JS, Regezi JA. Advanced diagnostic methods in oral and
       maxillofacial pathology. Part II: immunohistochemical and immunofluorescent methods. Oral Surg

Oral Med Oral Pathol Oral Radiol Endod 2002; 93(1): 56-74.
[http://dx.doi.org/10.1067/moe.2002.119567] [PMID: 11805778]

[20]   Leong AS, Wright J. The contribution of immunohistochemical staining in tumour diagnosis. Histopathology 1987; 11(12): 1295-305.
[http://dx.doi.org/10.1111/j.1365-2559.1987.tb01874.x] [PMID: 3326815]

[21]   Paraf A, Peltre G. Immunoassays in food and agriculture. Springer Science & Business Media 2012.

[22]   de Matos LL, Machado LN, Sugiyama MM, Bozzetti RM, da Silva Pinhal MA. Tecnologia aplicada na detecção de marcadores tumorais. ABC Medical Files 2005 Jul 25; 30(1)

CHAPTER 6

Protocols for the Detection and Proteome Analysis of the Yellow Mosaic Virus Infected Soyabean Leaves

Bapatla Kesava Pavan Kumar and **Surapathrudu Kanakala**[*]

Institute of Plant Sciences, Agricultural Research Organization, Volcani Center, Israel

Abstract: Soybean (Glycine max) is one of the legumes, susceptible to yellow mosaic disease caused by *Mungbean yellow mosaic India virus* (MYMIV) and *Mungbean yellow mosaic virus* (MYMV) infection. The quantitative proteomic analysis allows achieving deeper knowledge about the viral infection. For quantitative proteomic analysis, two-dimensional gel electrophoresis (2D-PAGE) is the common method of choice. Optimization is required even for the published protocols based on the type of sample to be analyzed and for the proteins of interest. We compared four different published protocols with some modifications and selected the one which is more effective in terms of resolution and reproducibility of 2D-PAGE. Here we present our simple and cost-effective procedure for the detection of viral infection and proteomic analysis of YMV infected soybean leaves without compromising the resolution and reproducibility of 2D-PAGE.

Keywords: Glycine max, Leaf proteome, Mungbean yellow mosaic India virus, Protein extraction, Soybean, Two-dimensional gel electrophoresis, Yellow mosaic disease.

INTRODUCTION

In India, *Mungbean yellow mosaic India virus* (MYMIV) and *Mungbean yellow mosaic virus* (MYMV) are the major causative agents of the yellow mosaic disease in legumes. These two viruses belong to the genus *Begomovirus* of the family *Geminiviridae* and are transmitted by whitefly (*Bemisia tabaci*) [1 - 3]. Begomoviruses cause extensive damage to a wide variety of crop plants.

Proteomics is the analysis of the complete functional protein complement of the genome under defined conditions. Comprehensive identification of proteins, their isoforms, as well as their prevalence in each tissue, characterizing the biochemical

[*] **Corresponding author Surapathrudu Kanakala:** Institute of Plant Sciences, Agricultural Research Organization, Volcani Center, Israel; E-mail: pksrivastava.bce@itbhu.ac.in

Sandeep Kumar & Dhiraj Kumar (Eds.)

and cellular functions of each protein and the analysis of protein regulation and its relation to other regulatory networks are the applications of proteomic approaches [4]. The standard method for proteome analysis involves high-resolution two-dimensional gel electrophoresis (isoelectric focusing/SDS-PAGE) (2DE) or gel-free liquid chromatography (LC) protein separation followed by mass spectrometric (MS) or tandem MS (MS/MS) identification of selected proteins [5 - 11].

Protein extraction and sample preparation are the most critical steps in any proteomics study [12]. Better reproducibility and accurate quantification can be achieved only with complete solubilization of proteins during extraction. In plant tissues, secondary metabolites, phenolic compounds, lipids, nucleic acids, cell wall and storage polysaccharides are the major contaminants, which affect the quality of the protein extraction and are responsible for poor resolution. A challenge still remaining in the accurate analysis of proteins is to get rid of contaminants in protein extraction that affect the performance of 2D-PAGE. In order to get high resolution in 2DE, proteins have to be completely denatured, disaggregated, reduced and solubilized to disrupt molecular interactions and to ensure that each spot represents an individual polypeptide.

The method of extraction of proteins from plants plays an important role as changes in the experimental conditions for extraction have been shown to pick up different proteins from the same plant tissue [13]. Thus optimization is mandatory even for the published methods based on the type of sample to be analyzed and for the proteins of interest. We compared four published methods [14 - 17] for total soluble protein extraction from diseased soybean leaves for proteomic analysis and optimized the conditions for sample preparation and 2D-PAGE. By introducing slight modifications, we made the current method cost-effective without compromising the resolution and reproducibility of 2D-PAGE [18]. Here we describe the protocols for simultaneous detection and proteomic analysis of YMV infected soybean leaves without compromising the resolution and reproducibility of 2D-PAGE.

METHOD

Protein Extraction and Virus Detection

1. Grind 500 mg of leaf tissue in liquid nitrogen using mortar and pestle.
2. Transfer ground tissue to 1 ml of extraction buffer (0.7 M Sucrose, 0.1 M KCl, 0.5 M Tris.HCl (pH 7.5), 50 mM EDTA, 1 mM PMSF, 50 mM DTT) in a 2 ml micro-centrifuge tube and vortex vigorously.
3. Directly use 0.5 μl from the extraction buffer to perform PCR using specific primers.

4. Centrifuge for 10 min at 20,000 g to clarify supernatant.
5. To the clarified supernatant, add 1 ml of Tris-Saturated Phenol (pH 8.0), and vortex vigorously at 4°C for 20-30 min and centrifuge at 20,000 g for 10 min.
6. Re-extract the phenolic phase with 1 ml of extraction buffer.
7. Add four volumes of 100% methanol containing 100 mM ammonium acetate and 10 mM DTT (chilled to -80°C) to the pooled phenolic fractions and incubate for 2-3 h at -80°C.
8. Centrifuge for 10 min at 20,000 g to precipitate proteins.
9. Wash protein pellet 2-3 times with 100% methanol containing 100 mM ammonium acetate and 10 mM DTT with centrifugation at 20,000 g for 10 min.
10. Repeat the wash with several volumes of ice-cold 90% ethanol with 10 mM DTT and centrifugation at 20,000 g for 10 min.
11. Dry pellet at room temperature and dissolve in resuspension buffer (8.5 M Urea, 2.5 M Thiourea, 5% CHAPS, 2% Triton X100, 100 mM DTT and 1% pH 3-10 Ampholytes) by vortexing at medium speed for 1 h at room temperature.
12. Perform sonication of 2 or 3 pulses 15 sec each at 22 W to improve dissolution.
13. After sonication clarifies protein solution by centrifugation at 20,000 g for 10 min at 18°C.

Iso-electric Focusing

1. Determine the concentration of protein by Bradford assay and use about 300 μg of total protein for loading.
2. Dilute protein samples in a HED buffer (2% 2-Hydroxyethyl disulfide (HED), 1% pH 3-10 Ampholytes, 5% Glycerol) to achieve a ratio of 3:1 of HED/DTT.,
3. Mix thoroughly and incubate at 4°C for 2 h prior to in-gel rehydration.
4. Apply samples on to an 18 cm, pH 3-10 or 4-7 linear or non-linear IPG strip and rehydrate passively for 16 h at 25°C.
5. Then focus the rehydrated strips at 18°C using Protean IEF Cell (Biorad) (500 V, Rapid ramp for 1 h; 1000 V, Rapid ramp for 1 h; 1000 – 8000 V, Linear ramp for 2 h; 8000 V, Rapid ramp for 46000 Vh).
6. Proceed to 2nd dimension immediately or the focused strips can be stored at -70°C until the sec dimension analysis.

SDS-PAGE/ 2nd Dimension

1. Perform initial reduction with 1% DTT in equilibration buffer for 15 min followed by alkylation with 2.5% IAA in equilibration buffer for 15 min.
2. Layer the strips on top of a 12.5% or 13% polyacrylamide gel and seal with 0.5% agarose in electrophoresis buffer.

3. Also, layer the wicks wetted with the molecular weight marker to the basic end of the strip for calibration.
4. Perform sec dimension electrophoresis initially at 5 W per gel for 30 min and then at 12 W per gel (constant power 600 V, and 60 mA per gel maximum) for 4-6 h.
5. After completion of electrophoresis, stain the gels with colloidal Coomassie staining [19].

Colloidal Coomassie Staining

1. After the completion of electrophoresis, dismantle the gel cassette and immerse the gel in fixing solution for 1 h.
2. Remove the fixing solution and wash the gel in deionized water for 10 min three times.
3. Place the gel in colloidal Coomassie blue staining solution and agitate slowly for overnight.
4. Destain the gel by repeated water washes until there was no background and document the gel.

Representative result of the two-dimensional gel electrophoresis of the healthy soybean leaf proteome (Fig. **1**) is reported.

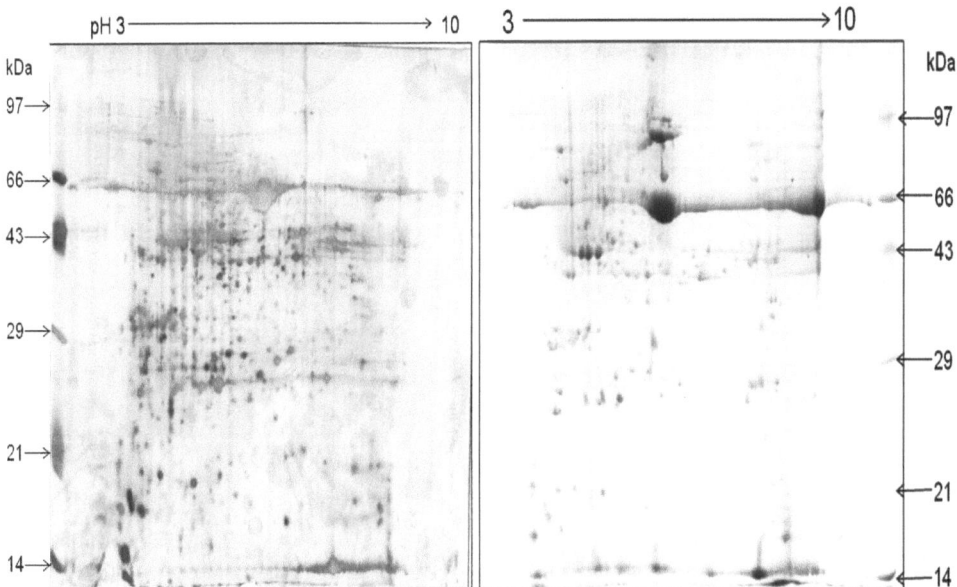

Fig. (1). Total protein from the healthy soybean leaves extracted using a modified phenol-based method and resolved on 2D gels. The gels were stained with colloidal Coomassie stain. Medium range molecular weight marker has been loaded for calibration.

Image Analysis

1. In order to get statistically significant data, at least two biological replicates with three experimental replicates from each sample are required for analysis.
2. Spot detection and quantification are performed in image analysis.
3. This can be done using any standard software for example PDQuest from Bio-Rad.
4. Perform quantitative analysis after normalization using total density in the gel image to determine the average quantity changes of the protein spots.
5. In order to level the scale of expression and to reduce the noise, log transform to base two the ratio of the average spot quantity.

CONSENT FOR PUBLICATION

Not applicable.

CONFLICT OF INTEREST

The author confirms that this chapter contents have no conflict of interest.

ACKNOWLEDGEMENTS

The authors acknowledge the Department of Biotechnology and the University Grants Commission, Government of India for funding and facilities. Dr. N. Kumaravadivel is acknowledged for providing the soybean samples.

ABBREVIATIONS

CHAPS	3-[(3-cholamidopropyl)dimethylammonio]-1-propanesulfonate;
DTT	Dithiothreitol;
EDTA	Ethylenediaminetetraacetic acid; Tris: tris(hydroxymethyl)amino-methane;
PMSF	phenylmethyl-sulfonyl fluoride;
SDS	Sodium dodecyl sulphate;
TCA	Trichloroacetic acid;

REFERENCES

[1]   Fauquet CM, Stanley J. Geminivirus classification and nomenclature; progress and problems. Ann Appl Biol 2003; 142: 165-89.
      [http://dx.doi.org/10.1111/j.1744-7348.2003.tb00241.x]

[2]   Girish KR, Usha R. Molecular characterization of two soybean-infecting begomoviruses from India and evidence for recombination among legume-infecting begomoviruses from South [corrected] South-East Asia. Virus Res 2005; 108(1-2): 167-76.
      [http://dx.doi.org/10.1016/j.virusres.2004.09.006] [PMID: 15681067]

[3]   Usharani KS, Surendranath B, Haq QMR, Malathi VG. Infectivity analysis of a soybean isolate of *Mungbean yellow mosaic India virus* by agroinoculation. J Gen Plant Pathol 2005; 71: 230-7.

[http://dx.doi.org/10.1007/s10327-005-0193-4]

[4] Wu DD, Hu X, Park EK, Wang X, Feng J, Wu X. Exploratory analysis of protein translation regulatory networks using hierarchical random graphs. BMC Bioinformatics 2010; 11(3) (Suppl. 3): S2.
[http://dx.doi.org/10.1186/1471-2105-11-S3-S2] [PMID: 20438649]

[5] Humphery-Smith I, Cordwell SJ, Blackstock WP. Proteome research: complementarity and limitations with respect to the RNA and DNA worlds. Electrophoresis 1997; 18(8): 1217-42.
[http://dx.doi.org/10.1002/elps.1150180804] [PMID: 9298643]

[6] Celis JE, Gromov P. 2D protein electrophoresis: can it be perfected? Curr Opin Biotechnol 1999; 10(1): 16-21.
[http://dx.doi.org/10.1016/S0958-1669(99)80004-4] [PMID: 10047502]

[7] Ong SE, Pandey A. An evaluation of the use of two-dimensional gel electrophoresis in proteomics. Biomol Eng 2001; 18(5): 195-205.
[http://dx.doi.org/10.1016/S1389-0344(01)00095-8] [PMID: 11911086]

[8] Garfin DE. Two-dimensional gel electrophoresis: an overview. Trac Trend Anal Chem 2003; 22: 263-72.
[http://dx.doi.org/10.1016/S0165-9936(03)00506-5]

[9] Carrette O, Burkhard PR, Sanchez JC, Hochstrasser DF. State-of-the-art two-dimensional gel electrophoresis: a key tool of proteomics research. Nat Protoc 2006; 1(2): 812-23.
[http://dx.doi.org/10.1038/nprot.2006.104] [PMID: 17406312]

[10] López JL. Two-dimensional electrophoresis in proteome expression analysis. J Chromatogr B Analyt Technol Biomed Life Sci 2007; 849(1-2): 190-202.
[http://dx.doi.org/10.1016/j.jchromb.2006.11.049] [PMID: 17188947]

[11] Penque D. Two-dimensional gel electrophoresis and mass spectrometry for biomarker discovery. Proteomics Clin Appl 2009; 3(2): 155-72.
[http://dx.doi.org/10.1002/prca.200800025] [PMID: 26238616]

[12] Rose JK, Bashir S, Giovannoni JJ, Jahn MM, Saravanan RS. Tackling the plant proteome: practical approaches, hurdles and experimental tools. Plant J 2004; 39(5): 715-33.
[http://dx.doi.org/10.1111/j.1365-313X.2004.02182.x] [PMID: 15315634]

[13] Saravanan RS, Rose JKC. A critical evaluation of sample extraction techniques for enhanced proteomic analysis of recalcitrant plant tissues. Proteomics 2004; 4(9): 2522-32.
[http://dx.doi.org/10.1002/pmic.200300789] [PMID: 15352226]

[14] Wang W, Scali M, Vignani R, *et al.* Protein extraction for two-dimensional electrophoresis from olive leaf, a plant tissue containing high levels of interfering compounds. Electrophoresis 2003; 24(14): 2369-75.
[http://dx.doi.org/10.1002/elps.200305500] [PMID: 12874872]

[15] Wang W, Vignani R, Scali M, Cresti M. A universal and rapid protocol for protein extraction from recalcitrant plant tissues for proteomic analysis. Electrophoresis 2006; 27(13): 2782-6.
[http://dx.doi.org/10.1002/elps.200500722] [PMID: 16732618]

[16] Natarajan S, Xu C, Caperna TJ, Garrett WM. Comparison of protein solubilization methods suitable for proteomic analysis of soybean seed proteins. Anal Biochem 2005; 342(2): 214-20.
[http://dx.doi.org/10.1016/j.ab.2005.04.046] [PMID: 15953580]

[17] Sarma AD, Oehrle NW, Emerich DW. Plant protein isolation and stabilization for enhanced resolution of two-dimensional polyacrylamide gel electrophoresis. Anal Biochem 2008; 379(2): 192-5.
[http://dx.doi.org/10.1016/j.ab.2008.04.047] [PMID: 18510937]

[18] Pavan Kumar BK, Kanakala S, Malathi VG, *et al.* Transcriptomic and proteomic analysis of the soybean plants infected with yellow mosaic virus. J Plant Biochem Biotechnol 2017; 26: 224.
[http://dx.doi.org/10.1007/s13562-016-0385-3]

[19]   Candiano G, Bruschi M, Musante L, *et al.* Blue silver: a very sensitive colloidal Coomassie G-250 staining for proteome analysis. Electrophoresis 2004; 25(9): 1327-33. [http://dx.doi.org/10.1002/elps.200305844] [PMID: 15174055]

CHAPTER 7

2D-DIGE A Powerful Tool for Proteome Analysis

Sudhir K. Shekhar[1], Jai Godheja[2] and Dinesh Raj Modi[1,*]

[1] Department of Biotechnology, Babasaheb Bhimrao Ambedkar University (A Central University), Vidya Vihar, Raibareilly Road, Lucknow-226025 (U.P.), India

[2] School of Life and Allied Sciences, ITM University, Atal Nagar, Raipur, Chhattisgarh-492001, India

Abstract: In the recent past, two dimensional gel electrophoresis has emerged as a powerful molecular biology tool for the comparative expression profiling of complex protein sample. It involves the separation as well as the resolution of diverse proteins sample on the basis of isoelectric points and molecular mass of protein in two dimension ways. In this way, it reflects the view of overall proteome status including differentiation in protein expression levels, post-translational modifications. *etc.* Moreover, this allows the identification of novel biological signatures, which may give a particular identity of pathological background to cells or tissues associated with various types of cancers and neurological disorders. Therefore, by utilizing such tools, one can clearly investigate and compare the effects of particular drugs on cells of tissues and also one can analyze the effects of disease on the basis of variations in protein expression profile at broad spectrum. Recently, to get more error-less and accurate proteome profile, conventional 2-D gel electrophoresis has been enhanced with the inclusion of different types of protein labeling dyes which enables a more comparative analysis of diverse protein sample in a single 2-D gel. In this advanced technique (2-D-DIGE), protein samples are labeled with three different types of CyDyes (Cy2, Cy3, and Cy5) separately and combined and further resolved on the same gel. This will facilitate the more accurate spot matching on a single gel platform and will also minimize the experimental variations as commonly reported in the conventional 2D-gel electrophoresis. Therefore, in the present proteomic research era, 2D-DIGE has proved to be an extremely powerful tool with great sensitivity for up to 125 ng of proteins in clinical research volubility especially, neurological and cancer related disorders.

Keywords: 2-D-DIGE, Cancer, Cy Dyes, Drugs, Electrophoresis, Molecular biology, Molecular mass, Neurological disorder, Proteome, Protein expression, Tissues.

* **Corresponding author Dinesh Raj Modi:** Department of Biotechnology, Babasaheb Bhimrao Ambedkar University (A Central University), Vidya Vihar, Raibareilly Road, Lucknow-226025 (U.P.), India;
E-mail: drmodilko@gmail.com

Sandeep Kumar & Dhiraj Kumar (Eds.)

INTRODUCTION

Two dimensional gel electrophoresis (2D-GE) is an important analytical method in various fields including proteome research [1]. This technique is highly reproducible and capable of generating statistically relevant data of cellular proteomes. Above all the results can be easily compared with each other with the help of appropriate analytical software. But still this method has a scope of improvement, a method originally introduced by Minden and colleagues [2] to overcome the inherent variability of 2D gels when comparing large data sets. A slight modification of 2DGE is 2D-DIGE (Two dimensional Difference Gel Electrophoresis) where fluorescent protein labels are used that has minimal effect on protein electrophoretic mobility and also allows the simultaneous electrophoresis of multiple samples on a single gel. This technique thus boosts up the reliability on results and authenticity is increased many a times.

DIGE is a technique where up to three different protein samples can be labelled with size-matched, charge-matched spectrally resolvable fluorescent dyes (for example Cy3, Cy5, Cy2) prior to two dimensional gel electrophoresis [2].

Methodology

Two dimensional Difference Gel Electrophoresis (2-D DIGE) is one of the very versatile and effective tools available in proteomics. It uses cyanine dye labeling of protein before proceeding for 2-D gel electrophoresis. Normally three cyanine dyes are used in this method which are Cy2, Cy3 and Cy5 [2 - 4]. These dyes contain an N-hydroxysuccinimidyl ester reactive group which forms a covalent bond with the epsilon amino group of lysine residues to yield an amide linkage. The lysine amino acid in proteins carries a positive charge. When these dyes are coupled to the lysine, replaces the lysine's single positive charge with its own, ensuring that the pI of the protein does not change (Fig. **1**).

Cyanine dye labeling gives a unique fluorescence property to proteins (Table **1**). When coupled to proteins, these dyes increase the molecular weight of the protein by 500 Da, however, as size and mass of all dyes are matched, a protein labeled by different dye will migrate to the same position.

A new strategy has recently emerged using new cyanine dyes that exhibit properties similar to the previously described dyes, however, instead of minimally labelling lysine residues they saturate cysteine residues. Again the cyanine dyes used are mass and charged matched; however, the saturation cyanine dyes are maleimide fluors that react with thiols, found on cysteine, with a nucleophilic addition to form a thioether. The reaction requires a pH range of 6.5-7.5, at which pH maleimides react faster with thiols than with amines, therefore they will preferentially label cysteine, however, there is a smaller chance of protein modification and pI shift [6]. A number of advantages come with this new strategy of labelling. Photobleaching is lower using both minimal and saturation Cy3 and Cy5 compared to a gel stained with Sypro Ruby [5] over a four hour time period. Saturation cysteine dyes were found to possess superior sensitivity, efficient enough to detect 0.1 ng of albumin compared to 1 ng by minimal lysine labelling, and almost double the number of spots detected using saturation Cy5 compared with minimal Cy5 [6]. As with the minimal lysine labelling cyanine dyes, if a cysteine amino acid is not present in a protein then the saturation dyes will not be able to label it, leaving the protein undectatable.

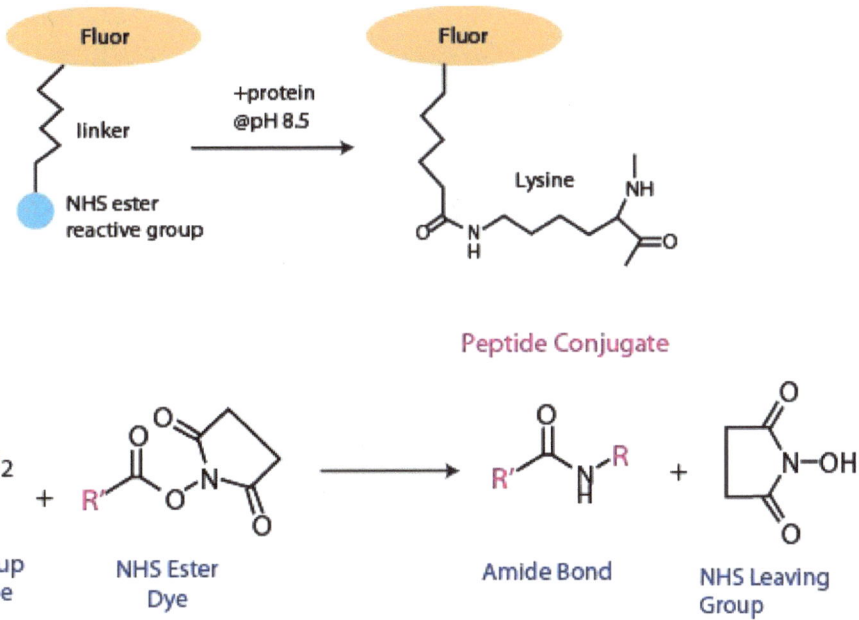

Fig. (1). Coupling of protein with minimal cyanine dye.

Table 1. Absorption and emission wavelength of fluorescent dyes.

Dye	Absorption wavelength (nm)	Emitted Wavelength (nm)
Cy2	488	520
Cy3	532	580
Cy5	633	670

When dye coupling is done, concentration of dye is kept limiting which leads to 1-2% of lysine residue labeling. In this process only one dye molecule per molecule is labeled. Even if multiples lysine residues of a protein are labeled, percentage of this double labeled species is too small to be visualized. As we have discussed that the binding of these dyes do not effect isoelectric point of protein but adds approximately Mr 500 mass. Unlabeled fraction of protein corresponding to a given spot resides at Mr 500 difference where fluorescence signal of the spot is seen. Once the spot of a protein of interest is identified by image analysis, recovery of the unlabeled protein by automated spot is achieved for various analysis.

Steps of the Experiment

Control is coupled with Cy3, sample 1 is coupled with Cy 5 and sample 2 is coupled with Cy 2. All three are then mixed and loaded for 2-D electrophoresis. After electrophoresis, the gel is scanned to get the following data: Cy3 fluorescent control image, Cy5 fluorescent sample 1 image and Cy2 fluorescent sample 2 image.

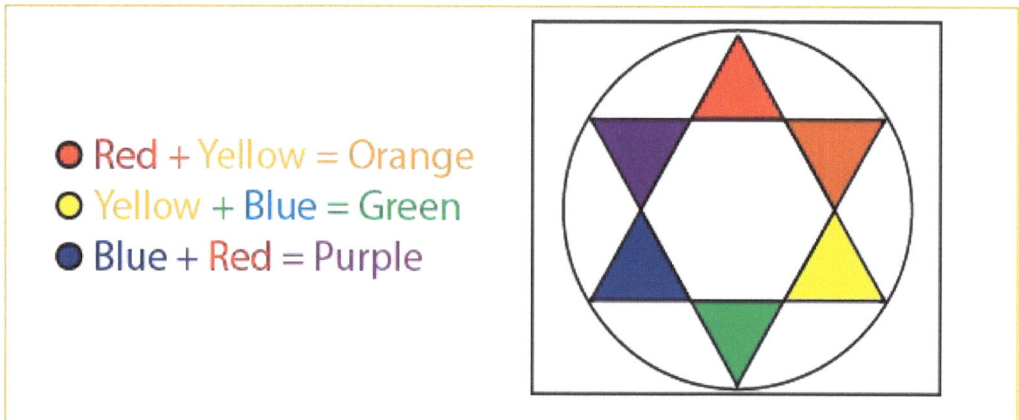

Fig. (2). Rules of color mixing.

Simple rule of color mixing (Fig. **2**) is followed in image analysis if expression of protein in control and sample 1 is equal the image overlay of Cy3 (blue) and Cy5 (red) will give purple spot. If these are difference in expression, image overlay will give different color. Similar control and sample 2 or sample 1 and sample 2 can also be analyzed for any change in proteome expression. This fluorescent method can detect even 125 pg protein [silver staining requires ng protein]. The Cy dyes technology, also called Fluorescence Difference Gel Electrophoresis (DIGE), is summarized in Fig. (**3**).

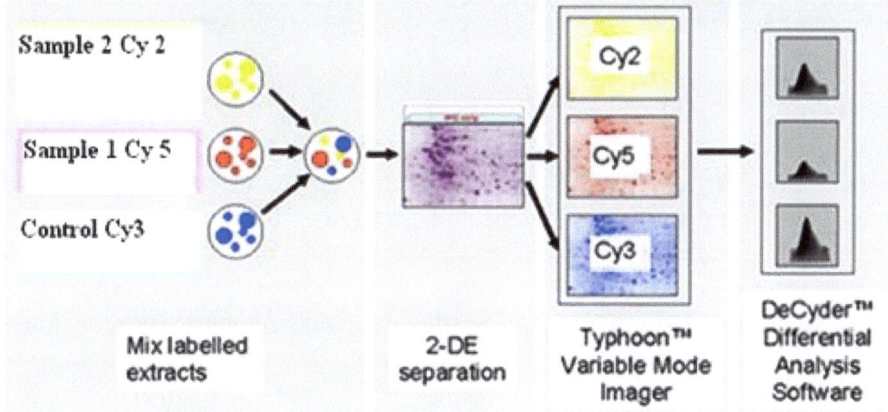

Fig. (3). Flow chart of DIGE.

ADVANTAGES OF DIGE

Multiplexing of samples in the same gel: Matching between different gels is more easy and straightforward. Further matching between images of the same gel is implicit because running differences are eliminated for the corresponding samples. Matching between gels can be performed between internal standard images, for which variations in spot patterns are due only to electrophoretic differences.

The use of the internal standard virtually eliminates experimental gel-to-gel variation and enables accurate quantification of induced biological change between samples. Quantitative comparisons of protein between samples are made based on the relative change of each protein spot to its own in-gel internal standard.

No technical replicates need to be run to confirm differences in protein abundance, thus reducing the number of gels required per experiment as well as analysis time and cost.

Sensitivity (sub nanogram level) and wide dynamic range (4-5 orders of magnitude) makes DIGE the method of choice

As low as 10% differential changes can be detected with high statistical accuracy

Applications of DIGE

Applications of 2D-DIGE can be found in virtually all research areas. With publications in the areas of cell signalling, developmental biology, plant proteomic analysis, neurosciences, clinical studies and different types of diseases including cancer.

CONCLUSION

Two dimensional Difference gel electrophoresis is becoming a high throughput technique for concluding the results in a short period of time. Through the use of the minimal lysine labelling 2DDIGE system, proteomics takes a step into the future. Employing innovative chemistry, this powerful tool enables direct comparisons of protein samples in a manner that was never previously possible. 2D-DIGE has already assisted in a better understanding of protein markers associated with infiltrating ductal carcinoma of the breast [10] the development of the cat brain [7, 8] the reaction of the Saccharomyces cerevisiae proteome to metal stress [9] and protein changes in mitochondrial proteins from mouse hearts deficient in creatine kinase [11], to name a few. Although DIGE is a relatively new technique it is continually developing into a better one, with the introduction of saturation cysteine dyes and development of internal standard controls. Armed with a technique of such potential, the possibilities for comparative proteomics expand beyond anything that was previously conceivable.

CONSENT FOR PUBLICATION

Not applicable.

CONFLICT OF INTEREST

The author confirms that this chapter contents have no conflict of interest.

ACKNOWLEDGEMENTS

Declared none.

REFERENCES

[1]   Sá-Correia I, Teixeira MC. 2D electrophoresis-based expression proteomics: a microbiologist's perspective. Expert Rev Proteomics 2010; 7(6): 943-53.
       [http://dx.doi.org/10.1586/epr.10.76] [PMID: 21142894]

[2]   Unlü M, Morgan ME, Minden JS. Difference gel electrophoresis: a single gel method for detecting changes in protein extracts. Electrophoresis 1997; 18(11): 2071-7.
       [http://dx.doi.org/10.1002/elps.1150181133] [PMID: 9420172]

[3]   Tonge R, Shaw J, Middleton B, *et al.* Validation and development of fluorescence two-dimensional differential gel electrophoresis proteomics technology. Proteomics 2001; 1(3): 377-96.

[http://dx.doi.org/10.1002/1615-9861(200103)1:3<377::AID-PROT377>3.0.CO;2-6]   [PMID: 11680884]

[4]   Alban A, David SO, Bjorkesten L, *et al.* A novel experimental design for comparative two-dimensional gel analysis: two-dimensional difference gel electrophoresis incorporating a pooled internal standard. Proteomics 2003; 3(1): 36-44.
[http://dx.doi.org/10.1002/pmic.200390006] [PMID: 12548632]

[5]   Gharbi S, Gaffney P, Yang A, *et al.* Evaluation of two-dimensional differential gel electrophoresis for proteomic expression analysis of a model breast cancer cell system. Mol Cell Proteomics 2002; 1(2): 91-8.
[http://dx.doi.org/10.1074/mcp.T100007-MCP200] [PMID: 12096126]

[6]   Shaw J, Rowlinson R, Nickson J, *et al.* Evaluation of saturation labelling two-dimensional difference gel electrophoresis fluorescent dyes. Proteomics 2003; 3(7): 1181-95.
[http://dx.doi.org/10.1002/pmic.200300439] [PMID: 12872219]

[7]   Van den Bergh G, Clerens S, Vandesande F, Arckens L. Reversed-phase high-performance liquid chromatography prefractionation prior to two-dimensional difference gel electrophoresis and mass spectrometry identifies new differentially expressed proteins between striate cortex of kitten and adult cat. Electrophoresis 2003; 24(9): 1471-81.
[http://dx.doi.org/10.1002/elps.200390189] [PMID: 12731035]

[8]   Van den Bergh G, Clerens S, Cnops L, Vandesande F, Arckens L. Fluorescent two-dimensional difference gel electrophoresis and mass spectrometry identify age-related protein expression differences for the primary visual cortex of kitten and adult cat. J Neurochem 2003; 85(1): 193-205.
[http://dx.doi.org/10.1046/j.1471-4159.2003.01668.x] [PMID: 12641741]

[9]   Hu Y, Wang G, Chen GY, Fu X, Yao SQ. Proteome analysis of Saccharomyces cerevisiae under metal stress by two-dimensional differential gel electrophoresis. Electrophoresis 2003; 24(9): 1458-70.
[http://dx.doi.org/10.1002/elps.200390188] [PMID: 12731034]

[10]   Kernec F, Unlü M, Labeikovsky W, Minden JS, Koretsky AP. Changes in the mitochondrial proteome from mouse hearts deficient in creatine kinase. Physiol Genomics 2001; 6(2): 117-28.
[http://dx.doi.org/10.1152/physiolgenomics.2001.6.2.117] [PMID: 11459927]

[11]   Kernec, F., Unlu, M., Labeikovsky, W., Minden, J.S., and Koretsky, A.P. (2001) Physiol Genomics 6: 117-28.
[http://dx.doi.org/10.1046/j.1471-4159.2003.01668.x] [PMID: 12641741]

CHAPTER 8

Molecular Techniques for Genotyping

Shalini Gupta[1,*], Somali Sanyal[2], Suresh Kumar Yadav[2] and **Madan Lal Brahma Bhatt[3]**

[1] *Department of Oral Pathology and Microbiology, King George's Medical University, Lucknow-226003, India*

[2] *Amity Institute of Biotechnology, Amity University, Uttar Pradesh, Lucknow-226028, India*

[3] *Department of Radiotherapy, King George's Medical University, Lucknow-226003, India*

Abstract: Genotyping is a process of determining the genetic constituent/genetic makeup "genotype" of an organism by examining the individual DNA sequence and comparing to a reference or other individual sequence. It helps the researchers to explore the genetic constitution, genetic linkages or variations like Single Nucleotide Polymorphisms (SNP) or multi-nucleotide changes in DNA. Identification of genotypes is also useful for determining their role in phenotypic expressions. Genotyping is an essential tool for researchers to find out disease-associated genes and gene variants. Genotype determined can also be used for the identification of susceptibility and prognosis for any disease and to find out responders/non-responders for a specific treatment, thus leading the way towards personalized medicine. Several molecular techniques have provided swift, reliable and accurate ways for determining genotypes. The process of genotyping involves molecular techniques like isolation and quantification of genomic DNA, visualization of DNA on agarose/polyacrylamide gel using electrophoresis, polymerase chain reaction (PCR), restriction fragment length polymorphism (RFLP), random amplified polymorphic detection (RAPD) of genomic DNA, amplified fragment length polymorphism (AFLP), sequencing, allele-specific oligonucleotide (ASO) probes, microarrays *etc*. The present chapter will describe the protocols for different molecular techniques that are used to determine genotypes.

Keywords: CTAB, DNA isolation, Genomic DNA, Genotype, Oligonucleotide, Plants, Protocol, PCR, Polymorphism, Polymerase chain reaction, RAPD, RFLP, SNP.

INTRODUCTION

In genomic disorders, advances in research had generated the quest to collect a large amount of good quality DNA from the sample. Recently, DNA typing is the

* **Corresponding author Shalini Gupta:** Department of Oral Pathology and Microbiology, King George's Medical University, Lucknow, U.P., India; E-mail: sgmds2002@yahoo.co.in

Sandeep Kumar & Dhiraj Kumar (Eds.)

most accurate method for the identification of human body fluid. Usual methods used to obtain genomic DNA are from saliva and peripheral blood [1]. Genetic analysis of common and prevalent diseases in large population is of great importance. This leads to the development of various safe, easy and economic techniques to obtain high quality genomic DNA. Though genomic DNA isolation from blood is a well established procedure to obtain high quantity and quality genomic DNA, yet they are not very desirable as it is an invasive techniques. And to obtain blood from masses is a herculean task. A non–invasive method *i.e.* DNA isolation from saliva provides an alternative to this massive task. Saliva has adequate number epithelial, Langerhans and leukocytes which have ample amount of genomic DNA.

DNA obtained through non-invasive methods like from saliva have adequate DNA yields from buccal (cheek) cells and oral-rinse (mouthwash) samples for PCR reactions and various genotyping protocols for hereditary epidemiological investigations. However, saliva also contains salts, food residue, bacteria and viruses and these can degrade human genomic DNA very quickly. Moreover, they show a quick and elevated growth rate when the samples are stored at room temperature [2, 3].

The collection of DNA from blood is an invasive and expensive procedure than saliva. DNA is extracted from blood to generate cell lines for functional studies. Trained technician is required for the collection of blood. Various examinations had demonstrated that the amount and nature of DNA from spit is of constrained use with a normal of 4.3 x 105 cells for each millilitre [4].

[1] Steps involved in DNA extraction from saliva:

1. Collection and storage,
2. Cell lysis,
3. RNase treatment,
4. Protein precipitation,
5. Ethanol precipitation,
6. DNA rehydration.

1. Saliva Collection and Storage

1. Before salivation gathering, the person's mouth ought to be free of nourishment or other outside material by flushing individual's mouth with water and not allowing the patient to eat or drink 30 minutes before gathering the saliva sample.

2. The Saliva sample is to be collected in 15 ml centrifuge tube having 2.5 ml of DNA adjustment buffer and 2.5 ml of saliva. The saliva is spit directly into the tube having DNA adjustment buffer.
3. Note: Collecting more than 2.5 ml of saliva will disturb the ratio of DNA and stabilization buffer, leading to the degradation of sample and less amount of saliva will give lower DNA yield from the expected.
4. The tube is capped once saliva is spitted in them and a homogenous mixture is obtained by mild mixing. These sample can be stored at room temperature (for shorter duration) or at 4°C (for duration >3 months).

2. Preparation for Cell Lysis

1. Ice box, pre warmed water bath (at 37°C)and 3 15 ml tubes are needed for DNA extraction. 3 tubes are required to hold cells, protein pellet and genomic DNA, respectively.
2. Vortexing of 2.5 ml of sample (having DNA and adjusting buffer) for 15 seconds followed by the addition of 5 ml cell lysis buffer to it and gentle mixing is to be done.
3. After mixing, this sample is incubated at room temperature for 30 minutes.

3. RNA Removal

1. To the above solution, 40 µl of RNAase solution (100 mg/ml) is to be added and followed by the incubation at 37°C for 15 minutes.
2. The solution is kept on ice for 3 minutes.
3. Increase the temperature of water bath to 65°C.

4. Protein and Lipid Removal

1. After keeping it on ice for 3 minutes, 50 µl Proteinase K solution (20 mg/ml) is added to this solution followed by gentle mixing and further incubating at room temperature for 30 minutes.
2. After incubation, 1.7 ml of protein precipitation solution is added to the mixtures and Vortexed for 20 seconds and again incubated on ice for 10 minutes.
3. After cooling on ice, the samples are centrifuged for 10 minutes again stored on ice to ensure a tight pellet.

5. Isolation and Purification of Genomic DNA

1. Decant the supernatant into a clean 15 ml tube. To this, 5 ml isopropanol and

    8µl of pure glycogen is added.

2. The mixture is gently mixed and centrifuged at 8000g, 4°C for 30 minutes. After centrifugation, the genomic DNA is precipitated. Discard the supernatant carefully. Rinse this pellet using 70% ethanol.
3. Centrifuge the sample for 1 min at 10,000xg and 20°C after initially washing.
4. This washing step can be repeated several times to ensure the good quality DNA.
5. Air dry the pellet.

6. Rehydration of g DNA

1. Add 300µl of Tris-EDTA to the dried sample to rehydrate the dried genomic DNA pellet.
2. Keeping it in a 65 °C hot water bath for 1 h. and over night at room temperature.

[2] Several Protocols Have Been Developed to Isolate DNA from Saliva

Common methods in use utilize polythene swabs or brushes, treated Guthrie-type cards which rinsed with saline or 3% sucrose solution. Feigelson *et al.* found that that saliva obtained after using a mouthwash containing alcohol is stable at room temperature for one week as it contains only about 40% bacteria in saliva.

Lum and Le Marchand suggested the use mouthwash (Listerine®, Johnson & Johnson) with alcohol for obtaining good quality DNA from saliva samples. They also demonstrated the reduced bacterial growth at 37°C for one week. This method is preferred for self-collection of samples. However, the only drawback associated to it is use of phenol-chloroform, which is already to be known as toxic and potent mutagenic and carcinogenic.

There are several kits available for more safe without using any toxic chemical for DNA extraction from saliva but they are quite costly, making it difficult to use them for a large population sized study. Thus search for more easy, safe and economic methods is still on. One of such procedure is as under.

Sample Collection

Patients are instructed to rinse their mouth throughly with 10 ml mouthwash (Listerine Cool Mint, Johnson & Johnson, containing 21.6% alcohol) for 30 seconds and spit the same in a 50 ml centrifuge tube.

DNA Extraction

- The saliva sample obtained from the patient is centrifuged at 11000g for 10 minutes at room temperature in order to separate mouthwash and the desired pellet.
- Discard the supernatant, and re-suspend the cellular pellet in 1 ml of PBS, vortex it for 30 seconds and transfer it to 2.5 ml microfuge tube.
- Centrifuge it for 5 minutes at 2000g, keep the pellet obtained and discard the supernatant.
- The pellet is again re-suspended in 1ml PBS and vortex for 30 seconds. Add 1 ml of cell-lysis buffer to it and vortex it for 30 seconds followed by incubation at 37°C for 10 minutes.
- To this solution, add 300µl of protein precipitation buffer and incubate this solution on ice for 5 minutes and centrifuge it at 15000g for 3 minutes.
- Decant the supernatant into a separate 2.5 ml microfuge tube and add 600µl of chilled isopropanol and mix it gently by inverting the tubes. Incubate at room temperature for 10 minutes followed by centrifugation at 15000g for 5 minutes.
- Discard the supernatant and wash the pellet using 600µl of freshly prepared 70% ethanol followed by centrifugation at 15000g for 5 minutes.
- Air-dry the pellets and dissolve them in 100µl of rehydration buffer. Keep the solution at 64°C for 1 hour and the it can be stored at -20°C for further use.

DNA Evaluation

The concentration of obtained DNA is determined using a spectrophotometer by scanning it at 260nm and calculating 260/280 ratio. DNA was separated using 0.8% agarose gel containing ETBr(0.5µg/ml) by electrophoresis and was visualized under UV light.

In order to get good saliva sample for genomic DNA extraction, following this can be done.

- Avoid high sugary foods/caffeine beverages immediately before sample collection, as they lower the pH and trigger the bacterial growth. For this reason a patient asked to rinse his/her mouth thoroughly with water to remove any food items present in the mouth at-least 30 minutes the before the sample collection and he/she is not allowed to eat or drink during this period.
- Any oral disease or injury should be well documented (if any).
- Passive drool technique can used to collect saliva, that give adequate amount of DNA. Buccal cells are collected by rubbing inside of cheeks for 1 minute with moderate pressure.
- The samples obtained must be quickly stored at 4°C and for longer duration of

storage, they can be initially stored at -20°C and finally transferred to -80°C. In -80°C the samples are stable for a period of 4 months. They can also be stored in swabs in which the samples are stable for 6 months.

Methods for DNA Extraction from Whole Blood

Isolation of genomic DNA from whole blood involves cell lysis, removal of protein and DNA Extraction [5].

The method can be either of these:

Two-step Lysis

1. Detergents like sodium dodecyl sulphate (SDS) and Triton™ X-100 are used to lyse RBC.
2. White blood cells lysis liberate the cell nucleus, genomic DNA and RNA.

One-step Lysis

- In the one-step lysis the red and white blood cells are lysed.
- The white cell nuclei and mitochondria are treated with a denaturation buffer.
- It contains a chaotropic salt and a protease which simultaneously denatures and digests the proteins in the cell nucleus and mitochondrion.
- Centrifugation is done to release the nuclei and mitochondria from the white blood cells, leaving the RNA in the supernatant.
- This leaves digested proteins and the genomic DNA in solution.

Advantages of the one-step lysis over the traditional two-step lysis method:

- DNA yield is increased and eliminates possible sample loss.
- No tube is changed.
- It saves time in the lab because of no overnight incubations and few steps are included.
- The one-step lysis approach is cheap as it doesn't require extra enzymes, like RNase, or any additional equipment.

This method not only quickly separates DNA from rest of the substances that hinder the downstream process but also separates it from any externally added chemicals.

Isolating Genomic DNA Using Magnetic Beads

- It is the innovative method that binds DNA on magnetic beads which are coated with silica.

- Whole blood cells are lysed by mild detergents like SDS, *etc.* This lysis of cells facilitates binding of DNA to the silica coated magnetic beads.
- Washing is done under magnetic field and un-binded substances are washed away after several rounds of washing.
- Magnetic bead binded DNA is eluted by using low salts buffers.

Using magnetic beads is advantageous as it is cheap, gives very good yield of DNA. The only drawback is that its setup is very expensive as it requires magnetic capture stands, plates and magnetic beads.

GENOMIC DNA ISOLATION OF DNA FROM BLOOD

DNA will be isolated with phenol-chloroform method.

Reagents Required

RBC Lysis buffer, Proteinase K, 10% SDS, PCI, Isopropanol.

Procedure

1. Add 500 μl of RBC lysis buffer to 500 μl of blood.
2. Centrifuge at 12000rpm for 5 min at 4°C.
3. Decant supernatant.
4. Re-suspend the pallet in triple distilled water.
5. Centrifuge at 12000rpm for 5 min at 4°C.
6. Decant supernatant.
7. Add 100 μl of Proteinase K and 2-3 drops of 10% SDS. Re-suspend the pallet.
8. Add 100 μl of 5N NaCl, 250 μl of triple distilled water and 400 μl of saturated phenol and 100 μl chloroform.
9. Centrifuge at 12000rpm for 15 min at 4°C.
10. Transfer aqueous layer into a new tube.
11. Add 1 ml of Phenol: Chloroform: Isoamyl alcohol(25:24:1).
12. Incubate for 10 min at 37°C.
13. Centrifuge at 12000rpm for 10 min at 4°C.
14. Transfer aqueous layer into a fresh tube.
15. Add 1 ml of chilled isopropanol and incubate it for 10 min at 37°C.
16. Centrifuge at 12000rpm for 10 min at 4°C.
17. Decant supernatant.
18. Air dry the pellet.
19. Dissolve the pellet in triple distilled water.
20. Dissolved DNA can be quantified by spectrophotometer or by flourometer qubit.
21. The qualitative analysis can be done by running DNA on 0.8% agarose gel

using electrophoresis.

VISUALIZATION OF DNA THROUGH AGAROSE GEL ELECTRO-PHORESIS

Materials Required

Agarose solution, DNA staining solution (Ethidium Bromide), Electrophoresis (running) buffer (TAE/TBE), 6x Gel-loading buffer, DNA Ladder, Test DNA.

METHOD

1. Take the clean and dry glass/acrylic gel casting tray and seal its edges with adhesive tape to prepare a mold. Mold is set on a horizontal segment of the bench. (Note: Some casting trays do not require sealing with tape).
2. Electrophoresis/running buffer (usually 1x TAE or 0.5x TBE) is prepared to fill the electrophoresis tank and to cast the gel.
3. Electrophoresis/running buffer solution of agarose of desired concentration depending upon the expected fragment size in the DNA sample(s) by adding adequate amount of agarose powder to a measured quantity of electrophoresis/running buffer in glass bottle and melting it until it becomes transparent.
4. Care should be taken to keep the cap of glass bottle slightly loose while melting the agarose solution either in water bath or in microwave. Time of melting agarose should be minimal.

Range of separation of linear DNA of different get concentrations

Agarose Concentration	Range of Separation
(% [w/v])	Molecules weight (kb)
0.3	5-60
0.6	1-20
0.7	0.8-10
0.9	0.5-7
1.2	0.4-6
1.5	0-2-3
2.0	0.1-2

1. Cool down the molten agar to 55°C and then add ethidium bromide (EtBr) in concentration of 0.5µg/ml and swirling is done to mix the solution thoroughly and gently, avoid air bubble formation.Precaution should be taken while handling EtBr, as it is carcinogenic thus gloves should be worn while handling it.

2. Pour this molten agarose solution at approx. 45°C into the mold/casting plate having a reasonable look over for framing the wells in the gel. Ensure 0.5-1.0 mm clearance between the comb and the plate with the goal that entire wells are framed.Gel should be of 3-5mm thickness. Evacuate any air rise in the gel or between/under the teeth's of the brush.

3. Allow molten agarose to solidify and set completely. Pour small quantity of electrophoresis/running buffer over the gel such that a thin layer is formed over the gel and expel the brush carefully. Pour off the electrophoresis/running buffer after removing the comb in the electrophoretic tank and remove the sealing tape carefully. Comb should be removed vertically, failing to which the wells may crack.

4. Move the gel in the electrophoretic tank and fill it with electrophoresis/running buffer such that it just submerges the gel (approx. 1 mm height of buffer over the gel).

5. Mix the DNA with 6x loading dye in ratio of 5:1, mix it gently by pipetting and then slowly and carefully load it into the well using a micropipette. Avoid formation of air bubble during mixing and add the mixture very slowly into the well to prevent spillage of samples from the wells. Minimum amount of DNA detected by EtBr is approx. 2 ng in a 5 mm wide DNA band. More sensitive SYBR GOLD can be used to detect even 20 pg of DNA in a band.

6. Cover the electrophoretic tank with the lid and connect the electrodes and switch it on. Low voltage and high migration time should be preferred over high voltage and less migration time to prevent gel melting, DNA damage and smiling of the bands.

7. Once bromophenol blue (tracking dye) of loading dye had covered 3/4th distance of the gel, switch off the power supply and then it can be visualized directly on trans-illuminator or using a gel documentation system.

8. Place the gel on trans-illuminator or on gel documentation system (preferably without the casting tray). EtBr will fluorescence in presence of a UV light. Most of the trans-illuminators emit 302 nm wavelength in which EtBR-DNA complex shows maximum absorbance.

9. Post soaking technique may also be done in which the gel is soaked in solution of Etbr in running buffer (0.5 µg/ml) for 30-45 min taken after by soaking in water for 20 min and then visualizing it of trans-illuminator/gel documentation system.

Reagents

6x Gel-loading Buffer

- 0.25% (w/v) bromophenol blue; 0.25% (w/v) xylene cyanol FF; 30% (v/v) glycerol in H_2O.
- Store at 4°C.

DNA Staining Solution (10mg/ml)

- Dissolve 100 mg ethidium bromide in 10 ml of H_2O and mix is thoroughly so that dye is completely dissolved. Store the solution in dark bottle or wrap the bottle by aluminum foil and keep it at room temperature.
- Care should be taken while handling EtBr as it a mutagen.

EDTA (0.5M pH 8)

- Dissolve 186.1 g of $Na_2EDTA.2H_2O$ in 800 ml of H_2O. Stir it on a magnetic stirrer. Adjust the pH to 8.0with NaOHpellets. The disodium salt of EDTA will not dissolve unless the pH of the solution is adjusted to approx. 8.0.

TAE (50X)

- Dissolve 242 g of Tris base, 57.1 ml of glacial acetic acid and 100 ml of 0.5 M EDTA (pH 8.0) in water to make final volume 1 liter.
- Final working solution (1X) is 40 mMTris-acetate and 1 mM EDTA.

TBE (5X)

- Dissolve 54 g of Tris base, 27.5 g of boric acid and 20 ml of 0.5 M EDTA (pH 8.0) in water to make final volume 1 liter.
- Final working solution (0.5X) is 45mMTris-borate and 1 mM EDTA.

TPE (10X)

- Dissolve 108 g Tris base, 15.5 ml of 85% (1.679 g/ml) phosphoric acid and 40 ml of 0.5 M EDTA (pH 8.0) in water to make final volume 1 liter.
- Final working solution (1X) 90mM Tris-phosphate and 2mM EDTA.

POLYACRYLAMIDE GEL ELECTROPHORESIS

Materials

Buffers and Solutions

Acrylamide:bisacrylamide (29:1) (% w/v), Ammonium persulfate (10% w/v),

TEMED, TrisCl(0.5M pH 6.8 and 1.5M pH 8.8), distilled water.

METHOD

1. Wash the glass plates and spacers in with mild detergent solution and rinse them completely with tap water took after distilled water. Clean the glass plates and spacers with 70% ethanol and let it dry. Glass plates must be free from any spots to avoid development of air bubbles during gel casting.
2. Seal the assembly of glass plates with spacers either by using 1% agar or agarose.
    i. To assemble the cassette, place spacers parallel to the edges of the larger plate and place the notched plate over it.
    ii. Clip the sides of the plates using binder's clip.
    iii. Move the spacers slightly inside (1-2 mm) from the edges to create a furrow/groove like arrangement between the two glass plates. Ensure that the side spacers and lower spacers at 90 degree to each other. This will prevent the leakage of gel from the corners.
    iv. Pour molten agar solution through into this furrow/groove. Give an angle to the glass plate so that this molten gel runs across the furrow/groove and seals it completely.
    v. Extra care should be taken at corners, and multiple layers of agar solution can be applied at corners.
    vi. Sealing of gel can be tested by pouring sterile water into the sealed arrangement. If sealing is correct, remove the water by inverting the plates and use micropipette to aspire last drop of water.

There are several models of casting tray available in which there is no requirement of sealing.

Prepare resolving gel as mentioned below. (Below mentioned calculation is for 10 ml)

S. No.	Components	10.0% Gel	12.5% Gel
1.	1.5 M Tris-HCl pH 8.8	2.5 ml	2.5 ml
2.	Acrylamide + Bis- acrylamide (29:1) 30%	3.3 ml	4.16 ml
3.	Distilled water	4.120 ml	3.260 ml
4.	TEMED	30 µl	30 µl
5.	Ammonium persulfate 10%	50 µl	50 µl

4. Once resolving gel is prepared, pour it into the cassette very slowly. Care should be taken the air bubbles are not formed during the gel pouring.

5. Add little amount of distilled water so that a thin layer of water is formed over the resolving gel. This thin layer or water prevents oxidation at the surface and also smoothens the upper layer of the gel.

6. After the polymerization of gel remove the water with the help of blotting paper/micropipette.

7. Prepare stacking gel as mentioned below. (Below mentioned calculation is for 5 ml).

S. No.	Component	3% Gel
1.	Distilled water	3.210 ml
2.	Acrylamide + Bis- acrylamide (29:1) 30%	0.5 ml
3.	0.5 M Tris-HCl pH 6.8	1.25 ml
4.	TEMED	15 µl
5.	Ammonium persulfate 10%	25 µl

8. When the stacking gel is prepared, pour it over the resolving gel. And place the comb in it. Ensure at least 5-1 mm between comb and the resolving gel.

9. When stacking gel is solidified, outline the comb on the bigger plate, remove the bottom spacer and attach this plate assembly on the electrophoretic tank with the help of binder's clip. Put some running buffer in the lower well and while attaching the plates avoid formation of air bubble below the gel.

10. Pour some running buffer in the upper chamber and then remove the comb carefully without disturbing the wells.

11. Mix the DNA sample with 6X loading dye and load it slowly into the well with the help of fine tip and a micropipette.

12. Run this gel at low voltage n and high migration time (voltage is 1-8 volts/cm of the gel). Higher voltage may result in gel heating and denaturing of DNA. If Possible run the gel in cold conditions.

13. When loading dye had run sufficient distance, the power supply is switched off and plates removed and dismantle the glass plate assembly by removing the spacers and opening the glass plates using a spatula.

14. Make a small cut and the resolving gel towards the well number 1. Remove the stacking gel and place the resolving gel into a solution of running buffer and EtBr (0.5µg/ml) for 30 min.

15. Rinse this gel with water and then visualize this gel on trans-illuminator or on a gel documentation system.

POLYMERASE CHAIN REACTION

Materials Required

Sterile water, 10x Amplification buffer, $MgCl_2$, dNTP Mixture (20 mM) containing all four dNTPs (pH 8.0), Thermostable DNA polymerase, Primers (forward and reverse) (20 μM) in H_2O, Template DNA.

Method

In a PCR tube or in sterile 200 μl microfuge tube or the well of a sterile microtiter plate, mix in the following order for 10 μl reaction:

Reagent	Stock Concentration	Volume added (μl)	Final Concentration
Sterile H_2O	1X	5.3	0.5 X
PCR buffer	10X	1.1	1X
MgCl2	25 mM	0.8	2.0 mM
dNTPs	1.25 mM	0.9	0.11mM
Forward primer	5μM	0.6	0.3 μM
Reverse primer	20 μM	0.6	0.3 μM
Taq DNA polymerase	5U/μl	0.1	0.05U/μl
Template DNA	10 ng/μl	1.0	1μg/μl

1. The quantity of template DNA may vary like, in case of yeast, it is 10ng, bacterial 1ng, plasmid 10pg and mammalian DNA is upto 1μg. This variation is because of the complexities of DNA.
2. Spin down the samples in a table top centrifuge. And place the tubes in the PCR machine.Care should be taken that the PCR tubes are clean and they fitted properly in machine and their lid is closed tightly. If PCR machine is not installed with a heated lid, add a drop of mineral oil over the top of reaction mixture to prevent any evaporation.

Set the PCR program for amplification using primary denaturation, denaturation, annealing, elongation/polymerization and final elongation.

1. Pre –Denaturation 45 sec at 94°C
2. Denaturation 30 sec at 94°C
3. Annealing 30 sec at 55°C (annealing temperature for different primers has to be optimized)
1. Elongation 45 sec at 72°C
2. Final Elongation 1 min at 72°C
3. Storage at 4°C

Step 2 to step 4 are in cycle of 30-35 times. Time and annealing temperature has to be optimized for reaction depending upon the type primer used.

1. Take 5-10µl from the reaction mixture and analyze them through electrophoresis on 1%-2% agarose gel or on polyacrylamide gel for amplification. An appropriate amplification should give a bright DNA band of expected size on the gel.

RESTRICTION FRAGMENT LENGTH POLYMORPHISM (RFLP)

Reagents Required

Restriction enzyme, enzyme restriction buffer, sterile water, enzyme dilution buffer, PCR product/Genomic DNA.

Method

1. Take 8µl PCR product/amplified DNA or genomic DNA in a microfuge tube and add 1µl restriction enzyme buffer.
2. If PCR product to in higher quantity, less amount of PCR product and volume adjusted with sterile water can be used.
3. To this mixture, add 1µl (5-10 units) of restriction enzyme and incubate it for 1-2 hours at 37°C.
4. Enzyme should be diluted using enzyme dilution buffer to get suitable dilution. Spin down the reaction mixture in a centrifuge before incubation.
5. After proper incubation, analyse the restricted product on 1% - 2% agarose gel by electrophoresis.

CONSENT FOR PUBLICATION

Not applicable.

CONFLICT OF INTEREST

The author confirms that this chapter contents have no conflict of interest.

ACKNOWLEDGEMENTS

Declared none.

REFERENCES

[1]   Ghatak S, Muthukumaran RB, Nachimuthu SK. A simple method of genomic DNA extraction from human samples for PCR-RFLP analysis. J Biomol Tech 2013; 24(4): 224-31.
        [http://dx.doi.org/10.7171/jbt.13-2404-001] [PMID: 24294115]

[2]   Mendoza ÁC, Volante BB, Hernández ME, *et al.* Design of a protocol for obtaining genomic DNA from saliva using mouthwash: Samples taken from patients with periodontal disease. J Oral Biol Craniofac Res 2016; 6(2): 129-34.
        [http://dx.doi.org/10.1016/j.jobcr.2016.01.002] [PMID: 27195211]

[3]   Rogers NL, Cole SA, Lan HC, Crossa A, Demerath EW. New saliva DNA collection method compared to buccal cell collection techniques for epidemiological studies. Am J Hum Biol 2007; 19(3): 319-26.
        [http://dx.doi.org/10.1002/ajhb.20586] [PMID: 17421001]

[4]   Goode MR, Cheong SY, Li N, Ray WC, Bartlett CW. Collection and extraction of saliva DNA for next generation sequencing. Journal of visualized experiments: JoVE 2014; 1(90)
        [http://dx.doi.org/10.3791/51697]

[5]   Extraction Protocol DNA. Choosing Whole Blood DNA Isolation Method http://bitesizebio.com/30255/how-to-choose-your-method-for-whole-blood-dna-extraction/

CHAPTER 9

Sodium Bisulfite Conversion of Human Genome for DNA Methylation Studies

Aastha Mishra* and **Qadar Pasha**

CSIR-Institute of Genomics and Integrative Biology, Delhi, India

Abstract: The regulation of transcription and translation of a gene under a given environment is dependent on several factors and epigenetics is one such factor, responsible for the differential expression of several genes in health and in various diseases. DNA methylation, an important epigenetics mechanism has been shown to play a vital role in numerous cellular processes, and the abnormal patterns of methylation have been linked to the number of human diseases. CpG islands, a short stretch of DNA enriched with CpG sites in the 5' end of a gene, although remains unmethylated but tends to methylate aberrantly upon certain environmental exposures. The methylation of the promoter region bearing transcriptional start sites of those genes that encodes tumor suppressors such as tumor protein p53, retinoblastoma-associated protein 1, tumor protein p16, breast cancer 1 and many more result in the reduced expression of these genes and have been implicated in a large number of cancers like retinoblastoma, colon, lung and ovarian. A growing number of human diseases have been found to be associated with the aberrant DNA methylation. Hence, a deep insight into the individual's epigenetic profile is the need of the hour. Several approaches have been developed to map DNA methylation patterns genome-wide. Some of these approaches include enzymatic digestion with methylation-sensitive restriction enzymes, the capture of 5-mC by methylated DNA-binding proteins followed by next-generation sequencing and methyl-DNA immunoprecipitation followed by sequencing of precipitated fragments. However, this chapter is going to describe the most recommended method for studying DNA methylation pattern, the method based on bisulfite sequencing. The bisulfite treatment of DNA converts unmethylated cytosine(s) to uracil(s), which are subsequently amplified as Ts by PCR. Hence, the bisulfite-treated DNA has mutations specifically at unmethylated Cs that can be mapped by Next-Generation sequencing.

Keywords: Bisulfite conversion, CpG island, DNA methylation, Deep sequencing, Epigenetics, Genomics, Immunoprecipitation, Next-Generation Sequencing, PCR, Retinoblastoma, Restriction enzymes, Transcription, Unmethylated.

---

* **Corresponding author Aastha Mishra:** CSIR-Institute of Genomics and Integrative Biology, Delhi, India; E-mail: aastha0602@gmail.com

Sandeep Kumar & Dhiraj Kumar (Eds.)

INTRODUCTION

Out of several ways, the methylation of DNA is one of the most common ways of regulating gene expressions in eukaryotes [1]. DNA methylation, as the name suggests, is the addition of the methyl group at the cytosine bases of the dinucleotide CpGs of eukaryotic DNA. As a result, they are converted to 5-methylcytosine by *de novo* DNA methyltransferase (DNMT) enzymes such as DNMT1, DNMT3A and DNMT3B (Fig. **1**). In mammals, methylation is found sparsely but globally with the exception of CpG islands, where high CpG contents are found. Vertebrate CpG islands are short interspersed DNA sequences (>500 bp) generally found in the 5' end of the gene that deviate significantly from the average genomic pattern by being GC-rich (G+C percentage greater than 55%), CpG rich (observed CpG/expected CpG of 0.65) and predominantly non-methylated [2].

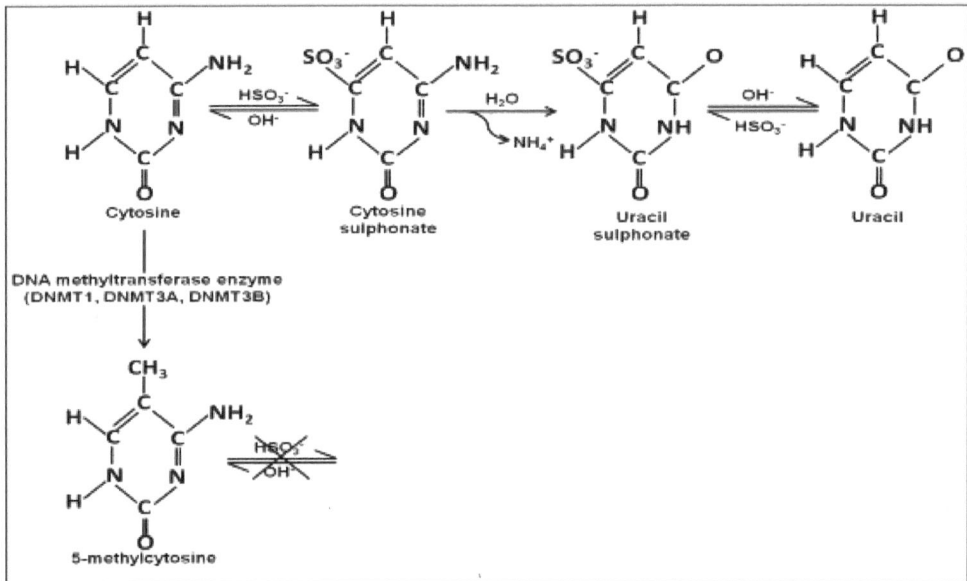

Fig. (1). Conversion of cytosine to 5-methylcytosine by DNA methyltransferase enzymes and action of sodium bisulfite on unmethylated cytosine residue. Bisulfite conversion changes unmethylated cytosine to uracil, while 5-methylcytosine remains unaffected by the treatment.

DNA methylation has emerged as an important process in numerous cellular processes like genomic imprinting, embryonic development, X-chromosome inactivation and many more (Fig. **2**). The first clue of the role of methylation in gene expression was provided by 5-azacytidine experiments in mouse studies [3].

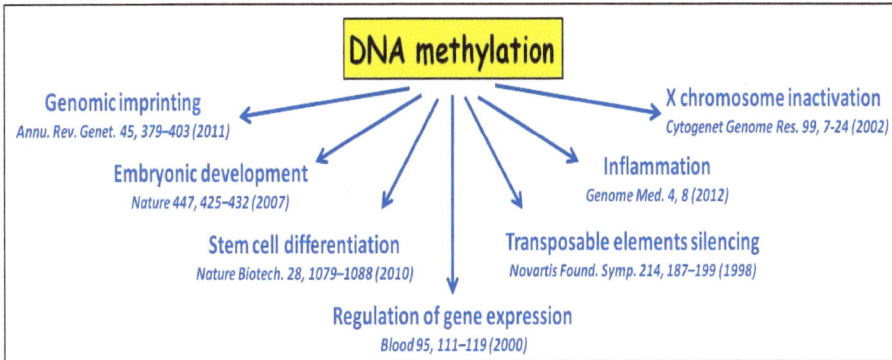

Fig. (2). DNA methylation affecting vital processes in the body.

The integration of 5-azacytidine in the growing strand of DNA severely inhibited the actions of DNMT enzymes to normally methylate DNA. Therefore, the comparisons of the cells before and after the treatment of 5-azacytidine allowed seeing what impact the loss of methylation had on gene expression [4]. The exact role of methylation in gene expression is unknown, perhaps it plays a crucial role in repressing gene expression by blocking the promoters at which activating transcription factors bind. Approximately, 70% of annotated gene promoters are associated with CpG islands, making it the most common promoter type in the genome [1]. Not all CpG islands found are associated with the promoter. Recent works have found a large class of islands that are remote from the transcription start sites (TSSs) but still show evidence for the promoter function. The lack of CpG dinucleotides in the vertebrate genome except the CpG islands is thought to be due to the loss of genomic CpGs due to deamination of methylated sequences [4].

Given the critical role of DNA methylation in gene expression and cell differentiation, it seems obvious that the errors in methylation could give rise to a number of devastating consequences, including various diseases. As a result, a growing number of human diseases have been found to be associated with aberrant DNA methylation [7]. The methylation of the promoter region bearing transcriptional start sites of many genes encoding tumor suppressors such as tumor protein p53, retinoblastoma-associated protein 1, tumor protein p16, breast cancer 1 and many more resulting in the reduced expression of these genes have been found in a large number of cancers like retinoblastoma, colon, lung and ovarian [7, 8]. 5-methylcytosine (5-mC) is spontaneously converted to thymine by deamination and is thought to be responsible for about one-third of all disease-causing mutations in the germline [4]. The mechanisms for establishing, maintaining and removing the methyl group are dependent on nucleosomal DNA and the histone modifications within the nucleosome [4]. Several approaches have been developed to map DNA methylation patterns on the genome. These methods include enzymatic digestion with methylation-sensitive restriction enzymes, the capture of 5-mC by methylated DNA-binding proteins followed by next-generation sequencing and methyl-DNA immunoprecipitation followed by sequencing of precipitated fragments [5]. However, the method based on the treatment of DNA with bisulfite is the most popular of all [6]. Bisulfite treatment converts unmethylated cytosine(s) to uracil(s) (Fig. **1**), which are subsequently amplified as Ts by PCR. This bisulfite-treated DNA is then sequenced to determine the percentage of C and T aligned to each C in genomic DNA sequence among all reads. The following section of this chapter will discuss in detail the method based on Bisulfite genomic sequencing for mapping DNA methylation.

Mapping Whole Genome DNA Methylation Using Sodium Bisulfite Conversion of DNA Followed by Next-generation Sequencing

Sodium Bisulfite Conversion of DNA

The bisulfite conversion of Genomic DNA (gDNA) selectively converts cytosine (C) to uracil (U) without significant transformation of 5-methylcytosine (5-mC) to thymine (T). This bisulfite converted DNA can then be processed further for their deep sequencing using various high throughput sequencing platforms for mapping the mutations as their methylation pattern. The recent advances in next-generation sequencing have made it possible to map DNA methylation at a higher resolution and in a larger number of samples. The mapping of specific regions of the genome could also be possible by PCR amplifying the desired sequences to be analyzed for assessing the methylation status in the samples (Fig. **3**). The bisulfite-conversion-based methylation PCR primers can be designed by the software methprimer (http://www.urogene.org/cgi-bin/methprimer/methprimer.cgi).

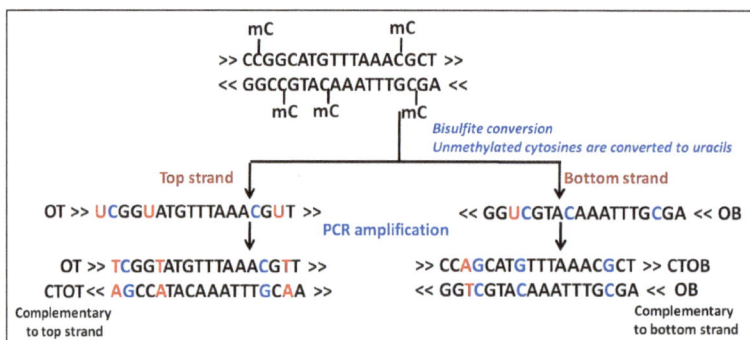

Fig. (3). Effect of bisulfite treatment to the DNA and their PCR amplification. Outline of bisulfite conversion of sample sequence of genomic DNA. Nucleotides in red are unmethylated cytosines converted to uracils by bisulfite, while blue nucleotides are 5-methylcytosines resistant to conversion.

Bisulfite genomic sequencing is a widely used method in research and provides a qualitative, quantitative and efficient approach to identify 5-mC at single base-pair resolution; hence, out of several methods, it is considered as a gold-standard technology for detection of DNA methylation. There are several commercially available kits in the market that could be used for bisulfite conversion of genomic DNA. In this chapter, we shall discuss in details the EZ DNA Methylation-GoldTM Kit (Zymo Research Corp., USA). This is one of the most widely used kits that integrate DNA denaturation and bisulfite conversion processes into one step.

Bisulfite Conversion Protocol

Firstly incubates the whole genome DNA sample with CT conversion reagent provided in the kit.

The CT conversion reagent is prepared by adding the following reagents to the CT conversion reagent tube.

Reagent	Volume, μl
M-Dilution Buffer	300
M-Dissolving Buffer	50
H_2O	900
Total Volume	1250

1. The content of the tube is vortexed for 10 min.

2. 20μl of 1μg genomic DNA sample each is mixed with 130μl of the CT conversion reagent and is placed in a thermal cycler for the following incubation.

3. 98°C for 10 min

4. 64°C for 2.5 hrs

5. After performing the incubation, the sample is loaded into Zymo-spin™ IC column. To this 600μl of M-Binding buffer is added and mixed by inverting the column several times and centrifuged at 12,000 x g for 30 sec.

6. 100μl of M-Wash buffer is added to the column and centrifuged at 12,000 x g for 30 secs.

7. 200μl of M-Desulphonation buffer is added to the column and kept at room temperature for 20 min. After the incubation, it is centrifuged at 12,000 x g for 30

secs.

8. The column is then washed by adding 200µl of M-Wash buffer and centrifuged at 12,000 x g for 30 secs. This step is repeated once.

9. The column is then placed into a 1.5ml microcentrifuge tube. 10µl of M-Dilution buffer is added and centrifuged at 12,000 x g for 30 secs to elute the bisulfite converted DNA.

10. This bisulfite converted DNA can then be used for PCR amplification of the entire genome. The forward and reverse primers in this step can be used from the specific next-generation sequencing kits, which one wants to use. The selection of the sequencing platform depends on the data one would like to generate and on the samples. We can use from a wide range of commercially available kits by different platforms such as Illumina, Affymetrix, Ion, Pacific Biosciences and a few more. For example, the HiSeq x instrument by Illumina gives the greatest number of reads compared to all other instruments, designed for whole human genome sequencing. While the PGM 318 instrument by Ion gives a fewer number of reads and shorter read length. Such data is ideal for small genomes or targeted sequencing and it also has fast turnaround time when compared to other sequencers. More details on these Next-generation sequencers are beyond the scope of this chapter.

11. A Master Mix is prepared for all reaction tubes with the following components in the order shown:

Reagent	Volume(µl)
10X First-Strand Buffer	2
Forward primers (10 pM/µl, kit specific)	1
Reverse primer (10 pM/µl, kit specific)	1
dNTP Mix (2.5 mM)	1
AmpliTaq Gold DNA polymerase	0.5
DNAase/RNAase free water	16.5
Total Volume	22

12. The contents are mixed well by vortexing and spinning the tube briefly.

13. 22µl of the Master Mix is aliquoted into each reaction tube with 3µl of bisulfite converted DNA of each sample and mixed thoroughly by pipetting. The tube is briefly centrifuged and immediately kept in the thermal cycler with the following conditions.

Stage	Step	Temperature	Time	Cycle
Holding	Denature	95°C	10 min	1
Cycling	Denature Anneal Extend	95°C 60°C 72°C	30 sec 30 sec 30 sec	35
Holding	Final extension –	72°C 4°C	10 min 10 min	1

The generated amplicons can then be purified from PCR reaction by various DNA purification kits commercially available. The choice of kit there will be depending on the sizes of the amplicons. One can choose from the wide variety of the kits available in the market. After the amplicon purification, it is very critical and crucial to analyze their length and concentration using a good bioanalyzer instrument such as Agilent High Sensitivity DNA chip on Agilent 2100 (Agilent Technologies, USA). This is the most important step as it will analyse the quality of the sample material. The quality and quantity of DNA are important in the bisulfite experiments. The good quality amplicons will give accurate and reproducible DNA methylation profiles. Those selected after elaborate scrutiny only should proceed for their deep sequencing using any kind of sequencing platforms to achieve single base-pair resolutions and for mapping DNA methylation.

CONCLUDING REMARKS

The bisulfite genome sequencing experiments are not limited to the sequencing of the bisulfite converted DNA, the analyses and interpretation of DNA methylation data are equally important. In fact, the post-sequencing steps are much more important and elaborate procedure, which is beyond the scope of this chapter. All of such experiments should go through the essential steps of data processing and quality control for the derivation of accurate DNA methylation maps [9]. The statistical analysis of the observed differences between the samples is the next important step in such experiments that will help in reaching precise conclusions and exact interpretations of the data. In conclusion, the bisulfite genomic sequencing of a good quality sample material could exhibit an accurate and reproducible DNA methylation profile that could give us a better assessment of individual's genomics and epigenomics profile and their risk to various diseases.

CONSENT FOR PUBLICATION

Not applicable.

CONFLICT OF INTEREST

The author confirms that this chapter contents have no conflict of interest.

ACKNOWLEDGEMENTS

Declared none.

REFERENCES

[1]   Deaton AM, Bird A. CpG islands and the regulation of transcription. Genes Dev 2011; 25(10): 1010-22.
[http://dx.doi.org/10.1101/gad.2037511] [PMID: 21576262]

[2]   Bird AP. CpG-rich islands and the function of DNA methylation. Nature 1986; 321(6067): 209-13.
[http://dx.doi.org/10.1038/321209a0] [PMID: 2423876]

[3]   McGhee JD, Ginder GD. Specific DNA methylation sites in the vicinity of the chicken beta-globin genes. Nature 1979; 280(5721): 419-20.
[http://dx.doi.org/10.1038/280419a0] [PMID: 460418]

[4]   Han L, Su B, Li WH, Zhao Z. CpG island density and its correlations with genomic features in mammalian genomes. Genome Biol 2008; 9(5): R79.
[http://dx.doi.org/10.1186/gb-2008-9-5-r79] [PMID: 18477403]

[5]   Shen L, Waterland RA. Methods of DNA methylation analysis. Curr Opin Clin Nutr Metab Care 2007; 10(5): 576-81.
[http://dx.doi.org/10.1097/MCO.0b013e3282bf6f43] [PMID: 17693740]

[6]   Yang AS, Estécio MR, Doshi K, Kondo Y, Tajara EH, Issa JP. A simple method for estimating global DNA methylation using bisulfite PCR of repetitive DNA elements. Nucleic Acids Res 2004; 32(3)e38
[http://dx.doi.org/10.1093/nar/gnh032] [PMID: 14973332]

[7]   Robertson KD. DNA methylation and human disease. Nat Rev Genet 2005; 6(8): 597-610.
[http://dx.doi.org/10.1038/nrg1655] [PMID: 16136652]

[8]   Gopalakrishnan S, Van Emburgh BO, Robertson KD. DNA methylation in development and human disease. Mutat Res 2008; 647(1-2): 30-8.
[http://dx.doi.org/10.1016/j.mrfmmm.2008.08.006] [PMID: 18778722]

[9]   Bock C. Analysing and interpreting DNA methylation data. Nat Rev Genet 2012; 13(10): 705-19.
[http://dx.doi.org/10.1038/nrg3273] [PMID: 22986265]

CHAPTER 10

Chromatin Immunoprecipitation (ChIP)

Kavyanjali Sharma[1], Subash Chandra Sonkar[2] and Shakuntala Mahilkar[*,3]

[1] *Department of Pathology, Faculty of Medicine, Banaras Hindu University, Varanasi, U.P.-221005, India*

[2] *Department of Obstetrics and Gynecology, Vardhman Mahaveer Medical College and Safdarjung Hospital, Ansari Nagar New Delhi- 1100029, India*

[3] *Hepatitis Division, National Institute of Virology, Pune, Maharashtra-411021, India*

Abstract: Chromatin immunoprecipitation or ChIP is an excellent method of investigation of the specific protein interaction and its altered forms with DNA region. These interactions have a significant role in various cellular processes such as replication, transcription, DNA damage repair, genome stability, gene regulation and segregation at mitosis. This technique is therefore giving us power to study a variety of cellular mechanisms inside the cell in terms of protein-DNA interaction. As the name Chromatin immunoprecipitation suggests this method utilizes chromatin preparation from cells to selectively immune-precipitate the protein of interest to identify DNA sequence associated with it. Chromatin is an organized structure of eukaryotic DNA which contains double-stranded DNA wrapped around nucleosomes. ChIP has been extensively used to depict transcription factors, variants of histone, chromatin modifying enzymes, post-translational modification of histone on the genome. In the classical ChIP method, protein and DNA is irreversibly cross-linked by UV exposure followed by immunoprecipitation with specific antibodies, protein-DNA complex is then purified, treated with proteases and then analysis is done by the method of Southern blot or dot blot using a radio-labelled probe derived from the cloned DNA fragment of interest. Further, it was modified by using formaldehyde for reversible cross-linking of protein-DNA complex and polymerase chain reaction for the detection of fragments of precipitated DNA. ChIP is a cumbersome procedure to perform and present many limitations, for example it requires many cells. Therefore, many modifications and variations, have also developed with the time which enables us to simplify the procedure and widen its range of applications. This chapter provides a brief method for Chromatin immunoprecipitation (ChIP) and its applications.

Keywords: ChIP, CPG islands, DNA methylation, Genome, Gene regulation, Immunoprecipitation, Mitosis, Protein-DNA interaction, Polymerase chain reaction, Replication, Transcription.

[*] **Corresponding author Shakuntala Mahilkar:** National Institute of Virology, Pune, Maharashtra-411021, India; E-mail: shakuntalamahilkar@gmail.com

Sandeep Kumar & Dhiraj Kumar (Eds.)
All rights reserved-© 2020 Bentham Science Publishers

OVERVIEW OF CHIP ASSAY

Chromatin immunoprecipitation (ChIP) is a technique to investigate the interaction of specific protein or it's modified forms with DNA region [1]. Protein-DNA interaction plays a significant role in various cellular processes such as replication, transcription, DNA damage repair, genome stability, gene regulation and segregation at mitosis.

This technique, therefore, giving us power to study various cellular process inside the cell regarding Protein-DNA interaction [2, 3]. As the name Chromatin immunoprecipitation suggests, this method utilizes chromatin preparation from cells to selectively immune-precipitate the protein of interest to identify DNA sequence associated with it. Chromatin is an organized structure of eukaryotic DNA which contains double-stranded DNA wrapped around nucleosomes. Interaction between specific DNA sequences with a variety of nuclear factors as well as with histone is crucial in various biological processes. ChIP has become a method of choice to explore all these interactions and provides a better understanding. ChIP has been extensively used to depict transcription factors, variants of histone protein, chromatin modifying enzymes, post-translational modification of histone on the genome.

David Gilmour and John T. L. are known as the pioneer of ChIP technique. They determined *in-vivo* distribution of RNA polymerase in bacterial gene by covalently cross-linking the protein to DNA in bacterial cells using UV irradiation following the lysis of the cells and immunoprecipitation of the bacterial RNA polymerase and finally, the DNA that is covalently attached to the RNA polymerase in the precipitate is purified and assayed by hybridization using probes corresponding to different regions of known genes [4]. Later, they applied same method to study the association of eukaryotic RNA polymerase II on Drosophila heat shock genes [5, 6]. The method was further modified by Solomon and Varshavsky. They used formaldehyde as cross-linking agent instead of UV irradiation and monitored the interaction between Drosophila hsp70 genes with histone H4 and RNA polymerase II. Formaldehyde treatment was given to the cells before and after heat shock. DNA was then subjected to shearing or restriction digestion. Cross-linked protein-DNA complexes containing histone H4 or RNA polymerase II were immunoprecipitated using specific antibodies. The protein-DNA complexes in the immunoprecipitants were heated to reverse the covalent cross-links, DNA fragments then purified and analysis was done by slot blot or Southern blot [7]. Further, Hecht and Grunstein modified the method by using PCR for the detection to study interaction of SIR proteins in *Saccharomyces cerevisiae* [8]. Rundlett *et al.*, in 1998 added PCR detection method in ChIP and investigated the histone modifications at specific loci in *Saccharomyces cerevisiae* [9]. Almost simultaneously, the ChIP method was optimized for mammalian cells, initially cross-linking with UV and then with formaldehyde [10 - 12]. Currently, ChIP method has been improved with the addition of several advanced methods. For example, advanced PCR and real time PCR detection method take the place of Dot blots hybridization. More advanced method based on ultrahigh-throughput DNA sequencing called ChIPSeq provides a large-scale chromatin immunoprecipitation assay [13, 14]. Although the ChIP was emerged with cross-linking step with formaldehyde it can also be performed without cross-linking step (native ChIP or N-ChIP). N-ChIP method is suitable for stably associated protein with DNA during chromatin processing and immunoprecipitation [15].

This chapter provides general guidelines, description of experimental setup and conditions for ChIP, the protein-DNA complexes immunoprecipitation, which can be analysed by PCR, qPCR, direct DNA sequencing or DNA microarrays. If instruments are different from those which have been described here, specific optimization might be required while following the ChIP protocol. In addition, further optimization of the procedure may require for different cell or tissue samples that may express differing protein levels.

GENERAL DESCRIPTION OF CHIP PROTOCOL

ChIP method includes the stabilization of protein genome interaction, cell lysis, nuclear material extraction, fragmentation of chromatin material, pre-clearing of solution to reduce nonspecific precipitation then immunoprecipitation, elution of precipitated DNA-protein complexes and finally reversal of stabilize protein genome interaction and characterization of DNA and proteins. Each step described below is tedious and requires standardization in a lab to perform a successful ChIP protocol.

1. Cross-Linking

In this step, DNA protein interactions are covalently stabilized. Generally, Formaldehyde is used as a cross-linking agent but, other cross-linkers also can be used namely, disuccinimidyl glutarate (DSG), ethylane glycol bis succinimidyl succinate (EGS), dimethyl adipimidate dihydrochoride(DMA), suberic acid bis (N-hydroxysuccinimide ester(DSS). Formaldehyde cross-linking shows certain limitations. It does not maintain all types of protein-DNA interactions for example co-activators interactions and transcription factors in case of hyper-dynamic equilibrium. Other limitation of using formaldehyde is that it is a zero-length cross-linker. To Trap larger protein complexes with complex quaternary structure longer cross-linkers such as EGS (16.1 Å) or DSG (7.7 Å) can be used. Also, sometimes two step cross-linking is also employed for more robust and highly efficient cross-linking [16, 17]. Condition and timing for cross-linking should be optimized for ChIP protocol. As Cross-linking may adversely affect sonication process and also may responsible for epitope sequestering.

2. Cell Lysis

In this step cells are lysed to bring DNA-protein complexes into solution. This step is achieved by detergent and salt-based buffers. Generally, treatment with ChIP lysis buffer is used for the lysis. Removal of cytosolic part is also done in this step as DNA-protein complexes are present in nuclear fractions. Mechanical lysis of cells is not recommended for this step.

3. Shearing of DNA

Extracted nuclear part contains unbound nuclear protein, intact chromatin and the cross-linked complexes of protein–DNA. For analysis of interacted DNA-protein part, long genomic DNA are fragmented into small pieces generally 200-1000bp. DNA shearing is achieved by either sonication or by enzymatic digestion with micrococcal nuclease (MNase). Each process has its own limitations and advantages. Enzymatic activity is very crucial for shearing and differences may lead to varying results, otherwise DNA shearing is highly reproducible by enzymatic treatment. Sonication needs high optimization and cold temperature maintenance. The sonicated chromatin can be stored up to 3 months in liquid nitrogen at -80°C. Multiple freeze thaw cycle should be avoided. Size of fragmented DNA can be determined in 1 or 1.5% agarose gel electrophoresis.

4. Immunoprecipitation

In this step ChIP validated antibodies are used for the selection of target protein, their interactions and removal of all other protein components. Antibodies can be monoclonal, polyclonal or oligoclonal specific for the target protein. If antibodies for the target proteins are not available, fusion proteins *i.e.* target protein can be tagged with His, Myc, human influenza hemagglutinin (HA), GST, T7, or V5 and expressed in the biological samples. Antibodies specific for tag can be used for immunoprecipitation. After the precipitation reaction complex of the antibody-protein-DNA is purified using beads that binds to primary antibodies (antibodies used for immunoprecipitation reaction) *i.e.* Protein G, protein A, or protein A/G. Blocking of the beads are required before use to reduce nonspecific interactions or background. Generally herring sperm DNA and BSA is used as blocking agents.

5. Reversal of Cross-links and Characterization of DNA

It can be done through heat incubation and proteinase-K treatments. However, if protein-DNA interactions are tough, high salt treatment can also be used before proteinase treatment. The DNA is purified either by phenol-chloroform method or using DNA purification kits or columns.

6. Analysis of DNA

The final step is analysis of DNA. It can be done by various methods, including PCR, quantitative real-time PCR and amplification for ChIP on chip, sequencing or cloning techniques.

Requirements

- Sonicator with micro tip
- Ultrasonic bath
- Shaking heat block
- Refrigerated microcentrifuge
- Means for quantitative PCR
- Tube rotator or tumbler at 4°C

Reagents

- Chelex 100
- Proteinase K
- Protein A–Sepharose
- Formaldehyde (! CAUTION if inhaled ingested or absorbed through skin it is very toxic)
- PMSF (! CAUTION after reaction with water it can form flammable gases. After disposal down a drain flush trap well. If absorbed through skin or ingested, it is toxic)
- Leupeptin
- Sodium molybdate dihydrate ($Na_2MoO_4 \_ 2H_2O$) (! CAUTION Harmful if inhaled or ingested).
- Glycerophosphate
- Sodium fluoride (NaF) (! CAUTION Very toxic if inhaled or ingested).
- Sodium orthovanidate (Na_3VO_4)
- p-Nitrophenylphosphate di(tris) salt
- SYBR Green PCR Master Mix
- Non-immune IgG fraction

Reagent Setup

1. **IP buffer**: NaCl-150 mM, Tris-HCl (pH 7.5)-50 mM, EDTA-5 mM, NP-40 (0.5% vol/vol), Triton X-100 (1.0% vol/vol).

Note: Per 1 ml IP buffer, add the following immediately before use and keep on ice: 5 ml of 0.1 M PMSF in isopropanol (–20°C), 1 µl of 10 µg µl–1 leupeptin (aliquots at –20 1C). Add the following phosphatase inhibitors if necessary: 10 ml of 10 mM $Na_2MoO_4.2H_2O$ (4°C), 10 ml of 1M β glycerophosphate (4°C), 10 ml of 1 M NaF (4°C), 1 ml of 100 mM Na_3VO_4 (aliquots at –20°C), 13.84 mg of p-nitrophenylphosphate (–20°C).

2. 1 M glycine: Dissolve 18.8 g glycine in ddH_2O (may require gentle heating) and bring up to 250 ml with ddH_2O

3.10% (wt./vol) Chelex 100: Add 1 g Chelex 100 resin to water (Milli-Q or NANO pure) and bring up to a final volume of 10 ml. Store at room temperature, 20–25 °C. (Stored at room temperature, 20–25°C)

4.20 µg ml–1 proteinase K prepared in Mili Q water: Dissolve 0.1 gm in 5 ml water (Milli-Q or NANOpure), aliquot, and store at –20 °C.

Step by Step Protocol

1. Start with ~50-100 million cells for each experimental condition. Add 40 µl formaldehyde (37% wt/vol) per millilitre of cell culture, incubate at room temperature for 10-15 min.

Note: Final concentration may vary from 0.75 to 1.42%. The efficiency of chromatin shearing, and precipitation of specific antigen can be affected by two factors *i.e.* the time of cross-linking and concentration of formaldehyde. The shearing efficiency can be improved by shorter time (5–10 min) of cross-linking and lowering formaldehyde concentrations (1%, wt/vol) but this can reduce the yield of precipitated chromatin by lowering the crosslinking efficiency especially for proteins that do not directly bind to DNA.

2. Add final concentration of 125mM glycine for quenching cross-linking and incubate at room temperature for 5min. (141µl of 1M glycine per 1 ml of the medium).

3. For adherent cells remove the growth medium and wash them, cells with 10ml of cold 1X PBS. After washing, scrape the cells into 1X PBS and collect in a 15ml conical tube. Centrifuge at 1500rpm for 10 min at 4°C and discard the supernatant. Wash the cells twice with 10ml of cold 1X PBS.

4. For non-adherent cells, transfer the cells in 15 ml conical tube and centrifuge at 1500 rpm for 10 min at 4°C. Discard the supernatant and wash the cells twice with 10ml of cold 1X PBS.

Note: Cell pellets can be stored at –80°C for at least 1 year.

5. Re-suspend the cell pellet in 10ml of IP buffer containing protease inhibitor.

Mix by pipetting up and down several times in a micro-centrifuge tube and Pellet the cell nuclei by centrifuging the cells at 12000g at 4°C for 1min. Carefully aspirate the supernatant.

6. Wash the cell pellet containing cell nuclei by re-suspending in 1ml IP buffer with protease inhibitors, followed by centrifugation and removing supernatant carefully.

7. Proceed with the sonication to shear the chromatin with the washed and re-suspended pellet in 1ml IP buffer containing protease inhibitors per 10 million cells.

Note: Volumes above 1ml are not recommended as it can reduce sonication efficiency. Sonication condition and time must be optimized for the experiment depending upon the cell and tissue type and sonicator model. In brief; in 1.5ml tube sonicate using microtip with a sonicator, do not allow the sample foaming, this will decrease the efficiency. To avoid foaming keep the tip of the sonicator probe just few millimetres away from the bottom of the tube. Stop sonication if foaming occurs and the process is paused till the bubbles rise to the surface. Hold the sample in an ice water bath to avoid sample heating during sonication. Also, total time of sonication is divided into small rounds with hold on ice in between to avoid extra heating (usually four rounds of sonication of 15seconds with two minutes rest in between). These series of small rounds of short pulses is more efficient than a single long-time sonication. The length of sonication and the power output must be determined to get optimum sonication conditions. For optimization 50% of highest power output is set for the microtip and 10-15seconds of small pulses of sonication is given for two, four, six or eight rounds with rest of 2 minutes in between on ice. The shearing efficiency can be checked by gel analysis of size of DNA fragments. If increasing rounds of short pulses of sonication or increasing time of sonication is not enough to obtain the desired fragment size of DNA, power of input should be increase. Total round of sonication should be decrease during high power input to avoid excessive heating of samples. for example, three rounds of 15 pulses at 50% power input for yeast cells and 90% duty cycle using a Branson Sonifier 200 [1] and for animal tissue, around four to six rounds (15 seconds of per round) at 50% power input using Misonix 300 [18]. DNA fragment size should be 200-1000 bp. To check fragment size, total DNA must be extracted from an aliquot of sheared chromatin, reversal of the cross-links is done and purification is done for a small volume of chromatin (purification is described in steps 7 and 19 below) and run on 1% (wt./vol) agarose gel.

8. After sonication, centrifuge the lysate at high speed in a microcentrifuge for 15min at 4°C and transfer the supernatant to a fresh microcentrifuge tube which contains sheared chromatin.

Note: At this time the lysate must be aliquoted (around two million cells per aliquot) for use with several antibodies. These aliquots can be stored at −80°C for months.

9. An aliquot of sheared chromatin is taken to a new microcentrifuge tube for isolation of total DNA, for determination of shearing efficiency and as a control for input DNA used in precipitations.
10. Measure the DNA concentration (A260) of the chromatin using a spectrophotometer. For blank use IP buffer.

Note: Usually the chromatin concentration should be >750ng/µl. The A260/A280 ratio should be ~1.4-1.6.

11. Aliquot 100µg of chromatin per antibody to be used into microcentrifuge tubes.
12. Dilute the chromatin to a final volume of 300µl with IP buffer supplemented with protease inhibitors.
13. Add antibody to the sample and incubate in an ultrasonic water bath for 15min at 4°C. A long incubation at 4°C should be applied in case of unavailability of ultrasonic bath. In the conventional method, the times of incubation range from 1 to 12hr and should be determined based on observation for each antibody. Use of proper control (a mock IP) is critical at this step which can be an irrelevant antibody such as α-glutathione S-transferase as a control or use of same antibody pre-incubated with saturating amounts of its peptide specific for the epitope at room temperature for 30min. Alternatively use of non-immune IgG fraction from the same species in which the antibodies were produced. Incubation with beads without antibodies can also be used as mock IP. Amounts of antibodies should be optimized for each protocol and incubation period may also vary for different proteins.
14. Centrifuge at 12000g for 10min at 4°C to clear the chromatin.
15. In between the step 12 and 13 wash protein A-agarose beads (20µl per IP sample) 3-4 times by adding 1ml IP buffer, centrifuging (1,000–2,000g) for a few secs at 20–25°C and aspirating the supernatant to remove ethanol. Dilute beads 1:1 with IP buffer and make aliquots of 40µl in fresh microfuge tubes.
16. Transfer the clear chromatin (~ 90%) from step 13 to an aliquot of Protein A agarose beads. Incubate the tubes at 4°C for 45min on rotating platform (20-30rpm).

Note: Avoid picking any precipitated material while adding to beads as this may carry aggregated non-specific DNA-complexes and contaminate IP.

17. Centrifuge the tubes at 1000-2000g for few secs and discard the supernatant. Wash the beads 4-5 times by resuspending with 1 ml cold IP buffer without

inhibitors.

18. Add 100µl of 10% Chelex100 to the washed beads at RT mix by brief (10s) Vortexing and boil for 10min.

19. The total DNA aliquot taken in Step 8 is precipitated three times volumes of chilled ethanol and washing with 70% ethanol is done. Elute the dried pellet in 100ml 10% (wt/vol) Chelex100 suspension, boil it for 10min and continue processing in the same way as the IP samples.

20. Cool the samples and add 1µl of 20µg per ml Proteinase K in each sample vortex and incubate at 55°C for 30min on thermal mixer (Optional).

21. Centrifuge at 12000g for 1min at 4°C and carefully transfer supernatant to a fresh tube. Avoid transfer of any chelex resin.

22. Add 120µl of Mili-Q water to the beads, Vortex for few secs to mix and centrifuge at 12000g for 1min at 4°C. Transfer the supernatant to a fresh tube.

23. Pool the supernatant from step 20 and 21, mix and store at 20°C. This supernatant can be further used for detection with PCR reactions.

Note: the sample stored at this step can be used even after more than 20 times freezing and thawing over a month without loss of PCR signal.

Troubleshooting

Problem	Possible reason	Troubleshooting
No specific signals	• Antibody used is not appropriate. • Cross-linking is not proper • Cross-linking time is very low or high. • Masking of epitopes.	• Try a different Antibody or proper controls for Antibody should be included. • Use different cross-linkers or two step cross-linking[16]. • Variant cross-linking time should be tested. • Use polyclonal antibody instead of monoclonal antibody.
Low Chromatin yield	• Low sample amount • Inefficient cell lysis.	• Accurate counting of cells should be done or increase the sample size.
Too much background	• Inadequate Pre-clearing. • Contamination of buffers.	• Increase the pre-clearing time of the Protein-A beads or increase the Beads. • Poor quality resin. Change resin. • Run the PCR for control without chromatin template to find out the possible contamination.
Very low PCR products	• PCR reaction not worked • Inefficient PCR cycle.	• Include a genomic DNA control of same sample as a positive control. • Try different PCR cycles.

Limitation

• It is not possible to define a common guideline for ChIP procedure appropriate for all situation for different starting materials and situations.

• This technique requires large amounts of cells (~10million) or starting material.

- Very time consuming.
- Each step requires tedious and exhaustive optimization processes.
- A successful ChIP experiment depends on the development and validation of a highly specific antibody to the bound protein or modification. Even different lots of same antibody preparations may vary the antibody quality varies. In fact, various histone modifications may alter the antibodies quality.
- There has also need of large efforts to improve tools for correct and précised result interpretation of sequence data output from ChIP-seq experiments.

Variation

To understand working of cellular machinery the basic need is to study molecular interaction between various proteins and modified proteins with the genomic DNA. ChIP is one of the most advanced methods to study these interactions. Despite of several limitations it has emerged as a powerful tool to understand *in vivo* interactions of proteins with genomic DNA. To overcome limitations several modifications has been applied to the basic ChIP procedure. There are two general types of ChIP procedure depending on the aim of the experiment *i.e.* **NChIP** and **XChIP**. XChIP uses chromatin fixing or cross-linking and fragmentation and the NChIP employs native chromatin prepared by nuclease digestion of cell nuclei. Generally, the protein which strongly interacts with DNA like Histone proteins and some transcription factors NChIP is more appropriate. Each procedure has own advantages and disadvantages [19].

Other limitation of conventional ChIP is that it requires large numbers of cells, generally at least 10^7 cells and suitable only for experimental systems based on cultured cells having high number of cells. For small number of cells high efficiency ChIP assay have been developed *i.e.* **Carrier ChIP (CChIP)** that can be done as few as on 100 cells. It is based on use of carrier chromatin from an evolutionary distant species. It can be applicable for primary cell samples obtained such as biopsies of normal and diseased tissue or by flow sorting, where cell numbers may limit conventional ChIP procedure. A limitation is that multiple aliquots of cells require for multiple modifications of proteins which may or may not be identical [20, 21].

Alternatively a **quick and quantitative (Q^{2}ChIP)** assay appropriate for the immunoprecipitation of histone and transcription factor from chromatin utilizes 1000cells or as few as 100 cells. Cross-linking the DNA-protein in suspension containing butyrate, background chromatin removal through a tube shift after washes, and a combination of cross-link reversal, protein digestion, increases the antibody-bead to chromatin ratio and a single step DNA elution considerably improves ChIP efficiency and shorten the process [22].

Micro-ChIP and Nano-ChIP-seq are also applicable for small number of cells as low as 10000 cells without use of carrier chromatin [23, 24].

In effort to reduce the time of total procedure **Fast ChIP** protocol has been developed. It has included use of ultrasonic bath and cation-chelating resin (Chelex100)-based DNA isolation which reduces the total time for preparation of PCR-ready templates to 1hr and avoided use of phenol chloroform extraction method [25].

Another variation of ChIP is **on bead PCR** which uses ChIP material directly as template in the PCR [26]. The **ChIP-on-beads** assay has been developed for quantitative assessments of ChIP products in a high through put manner. This method is compatible for flow cytometric analysis [27].

To increase efficiency more simplified form is developed *i.e.* **Matrix ChIP**. It uses 96 well plate coated with antibodies immobilized protein. It also maintains immobilized antibodies in correct orientation which enhances its binding capacity up to 10 folds [28, 29].

A different approach **ChIP-HAP** utilizes hydroxyapetide chromatography before immunoprecipitation reactions. This approach concentrates pure nucleosomes for immunoprecipitation reactions also eliminates all other proteins which mask epitopes of histone protein and hinder interaction of histone protein with antibodies. This also gives information about specific histone post-translational modification present within a specific genomic location on nucleosome [30].

In the **sequential ChIP method (SeqChIP)** protein–DNA complexes of the cells are subjected to two sequential immunoprecipitation reactions with two different specific antibodies. Its main application is to determine whether two proteins can simultaneously co-occupy a stretch of DNA *in-vivo* or verify the presence of bivalent histone marks in the same chromatin region. SeqChIP can be qualitative and quantitative both. In quantitative seqChIP full co-occupancy, no co-occupancy and partial co-occupancy can be determined for two proteins [31].

In another variant **ChIP-Bisulphite sequencing (ChIP-BS)** immunoprecipitation of one protein (a transcription factor, a histone modification, or a chromatin modifier) is done and then the eluted DNA is subjected to modification with sodium bisulphite and sequenced to know the DNA methylation status of that DNA [32].

Most advance and high throughput assay include ChIP coupled with DNA microarrays also known as **ChIP on chip**. The entire spectrum of *in-vivo* DNA-binding sites for a given protein can be determined by this method [33]. In this method immunoprecipitated material is tagged with fluorescent dyes and hybridized to DNA microarrays containing large number of probes. It has several advantages over conventional method identification of histone modification and permits discovery of unanticipated sites of protein-binding DNA on a genome-wide basis instead of a limited number of loci. Time saving as it has optimized commercially available platform and parallel analysis of thousands of genes based on different binding distributions or behaviours and permits statistical comparisons between classes.

For comparison of different samples such as when comparing the controls with treated samples it is necessary to normalize sample to sample variation using standards. For this purpose, a constant genomic region is selected and analysed simultaneously as a reference for normalization or as internal control in **ChIP-qPCR** method. Somehow the chromatin is unpredictably affected by experimental procedure and often prevents the choice of a suitable reference region in lack of knowledge. To overcome this method called **ChIP with reference exogenous genome (ChIP-Rx)** is developed. This method includes use of a constant amount of reference or "spike-in" epigenome which is added on a per-cell from another species that distinguishable *in silico* from the chromatin of the organism of interest. This exogenous chromatin is immunoprecipitated by antibody and normalize the sample to sample variation [34, 35]. However, this method is suitable only when working with antibodies against conserved proteins such as histone. For less conserved antigen an alternative strategy is the use of a synthetic external reference. For this purpose, synthetic DNA sequence tagged with digoxygenin (DIG) and cross-linked to an anti-DIG antibody is used. This DNA-DIG– antibody complex as external spike-in reagent [36].

Another advanced method to overcome the limitation of experiment with large number of cells is **BarChIP (Barcoded high-throughput ChIP-Seq)**. The ChIP-Seq involves chromatin immunoprecipitation followed by high-throughput sequencing (Seq) which includes ligation of sequencing adapters to the purified ChIP DNA and 30- to 35-nucleotide sequence reads are generated. The Binding site is determined by mapping the sequence of the DNA to the reference genome [37, 38].

In an improvement to ChIP-seq called **ChIP-exo**, digestion of one DNA strand in 5' to 3' direction of immunoprecipitated chromatin is done with the help of lambda exonuclease until it encounters a cross-linked protein, so that exact bases bordering a DNA-bound protein (the 'stop bases') can be mapped [39]. A recent method **Chip nexus** is an improved ChIP-exo method which combines the standard ChIP-exo method with the iCLIP method of library preparation protocol to map RNA-protein interactions for improvement of the efficiency by which

DNA fragments are added into the library. Addition of unique randomized barcode to the adapter, which enables the monitoring of over-amplification. This ChIP-nexus is more efficient as it requires only one successful ligation per DNA fragment [40].

Paired-End Tag Sequencing (ChIA-PET) is another variation in which ligase is added to the ChIP DNA to create chimeric DNA fragments followed by restriction enzyme digestion and paired-end tag sequencing. It provides high-resolution interaction data genome-wide which involves a given DNA-binding protein [24].

Conventional ChIP method does not provide information that protein binds to specific class of repeats or all the repeats. In conventional analysis based on protein binding to a particular repeat, it is assumed that all the repetitive sequences are same and will associate with that particular protein. For example, RNA polymerase I efficiently transcribes all types of rRNA genes sharing same sequences but differentially expressed. This distinct transcriptional activity is associated with distinct chromatin feature showing heterochromatic structures including CpG methylation and silent histone modifications such as methylation of H3K9 and H4K20. A recent approach **ChIP-chop method** is a robust way to analyze and distinguish the epigenetic composition of silent and active rDNA chromatin in organisms that evolved the CpG methylation system. The ChIP-chop method is a combination of ChIP technique with the analysis of meCpG levels of DNA associated with a defined protein using methylation-sensitive restriction enzymes [41].

Single-cell ChIP-seq is still under development to overcome the limitation of low number of cells. However, the sensitivity can be improved and successfully applied for ChIP analysis of low cells numbers [42].

The choice of variation or specific type of ChIP method depends on the aim of the researcher. Despite all of the above variations, more researches are going on to improve and overcome the limitations of ChIP methodology.

APPLICATION

The ChIP technique provides a wide range of application and with the advancement to overcome the technical limitation, its horizon is still expanding. Although it is not possible to list all the application of ChIP method, we are giving a broad idea to explore the biological processes by using this technique. The ChIP method is a technique of choice to identify the involvement of transcription factors and histone modifications at specific sites on genome as well as identifying new binding sites for transcription factors and mapping the location of histone modifications and histone variants at the genome-wide level. The application of the ChIP technique can be used to identify multiple proteins associated with a specific region of the genome and to identify many regions of the genome associated with a protein [43 - 46]. The ChIP assay is suitable to define the spatial and temporal connection of a specific protein-DNA interaction. For example, it can be used to define the sequential order of recruitment of several protein factors to a gene promoter or to map the relative amount of a histone modification across an entire gene locus during gene activation [43, 44]. The ChIP assay can also be used to analyze binding of transcription factors, transcription co-factors, DNA replication factors and DNA repair proteins [47, 48]. Exploring the accurate *in-vivo* interactions and modifications decipher how genetic machinery works in a cell which further add our knowledge of fundamental processes as immune reactions, stress response, disease progression and treatment.

CONSENT FOR PUBLICATION

Not applicable.

CONFLICT OF INTEREST

The author confirms that this chapter contents have no conflict of interest.

ACKNOWLEDGEMENTS

Declared none.

REFERENCES

[1]   Kuo MH, Allis CD. *In vivo* cross-linking and immunoprecipitation for studying dynamic Protein:DNA associations in a chromatin environment. Methods 1999; 19(3): 425-33.
[http://dx.doi.org/10.1006/meth.1999.0879] [PMID: 10579938]

[2]   O'Neill LP, Turner BM. Histone H4 acetylation distinguishes coding regions of the human genome from heterochromatin in a differentiation-dependent but transcription-independent manner. EMBO J 1995; 14(16): 3946-57.
[http://dx.doi.org/10.1002/j.1460-2075.1995.tb00066.x] [PMID: 7664735]

[3]   O'Neill LP, Turner BM. Immunoprecipitation of chromatin. Methods Enzymol 1996; 274: 189-97.
[http://dx.doi.org/10.1016/S0076-6879(96)74017-X] [PMID: 8902805]

[4]   Gilmour DS, Lis JT. Detecting protein-DNA interactions *in vivo*: Distribution of RNA polymerase on specific bacterial genes. Proc Natl Acad Sci USA 1984; 81(14): 4275-9.
[http://dx.doi.org/10.1073/pnas.81.14.4275] [PMID: 6379641]

[5]   Gilmour DS, Lis JT. *In vivo* interactions of RNA polymerase II with genes of Drosophila melanogaster. Mol Cell Biol 1985; 5(8): 2009-18.
[http://dx.doi.org/10.1128/MCB.5.8.2009] [PMID: 3018544]

[6]   Gilmour DS, Lis JT. RNA polymerase II interacts with the promoter region of the noninduced hsp70 gene in Drosophila melanogaster cells. Mol Cell Biol 1986; 6(11): 3984-9.

[http://dx.doi.org/10.1128/MCB.6.11.3984] [PMID: 3099167]

[7]   Solomon MJ, Larsen PL, Varshavsky A. Mapping protein-DNA interactions *in vivo* with formaldehyde: evidence that histone H4 is retained on a highly transcribed gene. Cell 1988; 53(6): 937-47.
[http://dx.doi.org/10.1016/S0092-8674(88)90469-2] [PMID: 2454748]

[8]   Hecht A, Strahl-Bolsinger S, Grunstein M. Spreading of transcriptional repressor SIR3 from telomeric heterochromatin. Nature 1996; 383(6595): 92-6.
[http://dx.doi.org/10.1038/383092a0] [PMID: 8779721]

[9]   Rundlett SE, Carmen AA, Suka N, Turner BM, Grunstein M. Transcriptional repression by UME6 involves deacetylation of lysine 5 of histone H4 by RPD3. Nature 1998; 392(6678): 831-5.
[http://dx.doi.org/10.1038/33952] [PMID: 9572144]

[10]   Boyd KE, Farnham PJ. Myc *versus* USF: discrimination at the cad gene is determined by core promoter elements. Mol Cell Biol 1997; 17(5): 2529-37.
[http://dx.doi.org/10.1128/MCB.17.5.2529] [PMID: 9111322]

[11]   Wathelet MG, Lin CH, Parekh BS, Ronco LV, Howley PM, Maniatis T. Virus infection induces the assembly of coordinately activated transcription factors on the IFN-β enhancer *in vivo*. Mol Cell 1998; 1(4): 507-18.
[http://dx.doi.org/10.1016/S1097-2765(00)80051-9] [PMID: 9660935]

[12]   Parekh BS, Maniatis T. Virus infection leads to localized hyperacetylation of histones H3 and H4 at the IFN-β promoter. Mol Cell 1999; 3(1): 125-9.
[http://dx.doi.org/10.1016/S1097-2765(00)80181-1] [PMID: 10024886]

[13]   Mikkelsen TS, Ku M, Jaffe DB, *et al.* Genome-wide maps of chromatin state in pluripotent and lineage-committed cells. Nature 2007; 448(7153): 553-60.
[http://dx.doi.org/10.1038/nature06008] [PMID: 17603471]

[14]   Johnson DS, Mortazavi A, Myers RM, Wold B. Genome-wide mapping of *in vivo* protein-DNA interactions. Science 2007; 316(5830): 1497-502.
[http://dx.doi.org/10.1126/science.1141319] [PMID: 17540862]

[15]   O'Neill LP, Turner BM. Immunoprecipitation of native chromatin: NChIP. Methods 2003; 31(1): 76-82.
[http://dx.doi.org/10.1016/S1046-2023(03)00090-2] [PMID: 12893176]

[16]   Tian B, Yang J, Brasier AR. Two-step cross-linking for analysis of protein–chromatin interactions. Trans Regul Methods Protoc 2012; 105-20.
[http://dx.doi.org/10.1007/978-1-61779-376-9_7]

[17]   Nowak DE, Tian B, Brasier AR. Two-step cross-linking method for identification of NF-kappaB gene network by chromatin immunoprecipitation. Biotechniques 2005; 39(5): 715-25.
[http://dx.doi.org/10.2144/000112014] [PMID: 16315372]

[18]   Chaya D, Zaret KS. Sequential chromatin immunoprecipitation from animal tissues. Methods Enzymol 2004; 376: 361-72.
[http://dx.doi.org/10.1016/S0076-6879(03)76024-8] [PMID: 14975318]

[19]   Turner B. ChIP with native chromatin: advantages and problems relative to methods using cross-linked material.

[20]   O'Neill LP, VerMilyea MD, Turner BM. Epigenetic characterization of the early embryo with a chromatin immunoprecipitation protocol applicable to small cell populations. Nat Genet 2006; 38(7): 835-41.
[http://dx.doi.org/10.1038/ng1820] [PMID: 16767102]

[21]   Rodríguez-Ubreva J, Ballestar E. Chromatin immunoprecipitation. Func Analy DNA Chrom 2014; 309-18.

[22]   Dahl JA, Collas P. Q2ChIP, a quick and quantitative chromatin immunoprecipitation assay, unravels epigenetic dynamics of developmentally regulated genes in human carcinoma cells. Stem Cells 2007; 25(4): 1037-46.
[http://dx.doi.org/10.1634/stemcells.2006-0430] [PMID: 17272500]

[23]   Acevedo LG, Iniguez AL, Holster HL, Zhang X, Green R, Farnham PJ. Genome-scale ChIP-chip analysis using 10,000 human cells. Biotechniques 2007; 43(6): 791-7.
[http://dx.doi.org/10.2144/000112625] [PMID: 18251256]

[24]   Furey TS. ChIP-seq and beyond: new and improved methodologies to detect and characterize protein-DNA interactions. Nat Rev Genet 2012; 13(12): 840-52.
[http://dx.doi.org/10.1038/nrg3306] [PMID: 23090257]

[25]   Nelson JD, Denisenko O, Bomsztyk K. Protocol for the fast chromatin immunoprecipitation (ChIP) method. Nat Protoc 2006; 1(1): 179-85.
[http://dx.doi.org/10.1038/nprot.2006.27] [PMID: 17406230]

[26]   Kohzaki H, Murakami Y. Faster and easier chromatin immunoprecipitation assay with high sensitivity. Proteomics 2007; 7(1): 10-4.
[http://dx.doi.org/10.1002/pmic.200600283] [PMID: 17152093]

[27]   Székvölgyi L, Bálint BL, Imre L, *et al.* Chip-on-beads: flow-cytometric evaluation of chromatin immunoprecipitation. Cytometry A 2006; 69(10): 1086-91.
[http://dx.doi.org/10.1002/cyto.a.20325] [PMID: 16998871]

[28]   Peluso P, Wilson DS, Do D, *et al.* Optimizing antibody immobilization strategies for the construction of protein microarrays. Anal Biochem 2003; 312(2): 113-24.
[http://dx.doi.org/10.1016/S0003-2697(02)00442-6] [PMID: 12531195]

[29]   Collas P. The current state of chromatin immunoprecipitation. Mol Biotechnol 2010; 45(1): 87-100.
[http://dx.doi.org/10.1007/s12033-009-9239-8] [PMID: 20077036]

[30]   Brand M, Rampalli S, Chaturvedi CP, Dilworth FJ. Analysis of epigenetic modifications of chromatin at specific gene loci by native chromatin immunoprecipitation of nucleosomes isolated using hydroxyapatite chromatography. Nat Protoc 2008; 3(3): 398-409.
[http://dx.doi.org/10.1038/nprot.2008.8] [PMID: 18323811]

[31]   Geisberg JV, Struhl K. Quantitative sequential chromatin immunoprecipitation, a method for analyzing co-occupancy of proteins at genomic regions *in vivo*. Nucleic Acids Res 2004; 32(19)e151
[http://dx.doi.org/10.1093/nar/gnh148] [PMID: 15520460]

[32]   Matarazzo MR, Lembo F, Angrisano T, *et al. In vivo* analysis of DNA methylation patterns recognized by specific proteins: coupling CHIP and bisulfite analysis. Biotechniques 2004; 37(4): 666-668, 670, 672-673.
[http://dx.doi.org/10.2144/04374DD02] [PMID: 15517979]

[33]   Pillai S, Chellappan SP. ChIP on ChIP assays: genome-wide analysis of transcription factor binding and histone modifications. Chromatin Protocols: Sec Edition 2009; 341-66.
[http://dx.doi.org/10.1007/978-1-59745-190-1_23]

[34]   Bonhoure N, Bounova G, Bernasconi D, *et al.* Quantifying ChIP-seq data: a spiking method providing an internal reference for sample-to-sample normalization. Genome Res 2014; 24(7): 1157-68.
[http://dx.doi.org/10.1101/gr.168260.113] [PMID: 24709819]

[35]   Orlando DA, Chen MW, Brown VE, *et al.* Quantitative ChIP-Seq normalization reveals global modulation of the epigenome. Cell Rep 2014; 9(3): 1163-70.
[http://dx.doi.org/10.1016/j.celrep.2014.10.018] [PMID: 25437568]

[36]   Eberle AB, Böhm S, Östlund Farrants AK, Visa N. The use of a synthetic DNA-antibody complex as external reference for chromatin immunoprecipitation. Anal Biochem 2012; 426(2): 147-52.
[http://dx.doi.org/10.1016/j.ab.2012.04.020] [PMID: 22543092]

[37]   Raha D, Hong M, Snyder M. ChIP-Seq: a method for global identification of regulatory elements in the genome. Curr Protoc Mol Biol 2010; 1-14.
[http://dx.doi.org/10.1002/0471142727.mb2119s91] [PMID: 20583098]

[38]   Chabbert CD, Adjalley SH, Klaus B, *et al.* A high-throughput ChIP-Seq for large-scale chromatin studies. Mol Syst Biol 2015; 11(1): 777.
[http://dx.doi.org/10.15252/msb.20145776] [PMID: 25583149]

[39]   Rhee HS, Pugh BF. ChIP-exo method for identifying genomic location of DNA-binding proteins with near-single-nucleotide accuracy. Curr Protoc Mol Biol 2012; 24.
[PMID: 23026909]

[40]   He Q, Johnston J, Zeitlinger J. ChIP-nexus enables improved detection of *in vivo* transcription factor binding footprints. Nat Biotechnol 2015; 33(4): 395-401.
[http://dx.doi.org/10.1038/nbt.3121] [PMID: 25751057]

[41]   Santoro R. Analysis of chromatin composition of repetitive sequences: the ChIP Chop assay. Funct Anal DNA Chrom 2014; pp. 319-28.

[42]   Rotem A, Ram O, Shoresh N, *et al.* Single-cell ChIP-seq reveals cell subpopulations defined by chromatin state. Nat Biotechnol 2015; 33(11): 1165-72.
[http://dx.doi.org/10.1038/nbt.3383] [PMID: 26458175]

[43]   Agalioti T, Lomvardas S, Parekh B, Yie J, Maniatis T, Thanos D. Ordered recruitment of chromatin modifying and general transcription factors to the IFN-β promoter. Cell 2000; 103(4): 667-78.
[http://dx.doi.org/10.1016/S0092-8674(00)00169-0] [PMID: 11106736]

[44]   Soutoglou E, Talianidis I. Coordination of PIC assembly and chromatin remodeling during differentiation-induced gene activation. Science 2002; 295(5561): 1901-4.
[http://dx.doi.org/10.1126/science.1068356] [PMID: 11884757]

[45]   Mikkelsen TS, Ku M, Jaffe DB, *et al.* Genome-wide maps of chromatin state in pluripotent and lineage-committed cells. Nature 2007; 448(7153): 553-60.
[http://dx.doi.org/10.1038/nature06008] [PMID: 17603471]

[46]   Lee TI, Jenner RG, Boyer LA, *et al.* Control of developmental regulators by Polycomb in human embryonic stem cells. Cell 2006; 125(2): 301-13.
[http://dx.doi.org/10.1016/j.cell.2006.02.043] [PMID: 16630818]

[47]   Weinmann AS, Farnham PJ. Identification of unknown target genes of human transcription factors using chromatin immunoprecipitation. Methods 2002; 26(1): 37-47.
[http://dx.doi.org/10.1016/S1046-2023(02)00006-3] [PMID: 12054903]

[48]   Wells J, Farnham PJ. Characterizing transcription factor binding sites using formaldehyde crosslinking and immunoprecipitation. Methods 2002; 26(1): 48-56.
[http://dx.doi.org/10.1016/S1046-2023(02)00007-5] [PMID: 12054904]

CHAPTER 11

Osteosarcoma Cell Culture and Maintenance to Detect the Apoptotic Effect of Some Promising Compounds by Potent Markers *viz.* DNA Fragmentation and Caspase-3 Activation

Asif Jafri*, Juhi Rais, Sudhir Kumar and Md Arshad*

Molecular Endocrinology Lab, Department of Zoology, University of Lucknow, Lucknow, India

Abstract: Osteosarcoma is the most common type of malignancy of bone and generally occurs among adolescent and young adults. Like the osteoblast cells of normal bone, osteosarcoma also forms the bone matrix, but the osteoid is not as strong as that of normal bones. Osteosarcoma is characterized by the production of weak or immature bones by the malignant cells. As the diagnosis of osteosarcoma is extremely poor, it suggests a critical need to develop some promising anti-osteosarcoma drugs to improve disease outcome. Many anti-cancer compounds induce apoptotic cell suicide *via* some potent cellular, molecular and biochemical markers. The cancer cell lines provide a valuable model system to study an extensive variety of cancer characteristics in the cell biology, genetics and chemotherapy or the impact of therapeutic molecules. The methods presented in this chapter describe the experimental technique used to culture the osteosarcoma cells for the documentation of DNA fragmentation and Caspase-3 activation associated with apoptosis.

Keywords: Adolescence, Apoptosis, Cancer cell line, Caspase-3, Chemotherapy, DNA fragmentation, Malignancy, Molecular and biochemical markers, Osteosarcoma, Therapeutic molecules.

INTRODUCTION

Osteosarcoma or the tumour of bone normally develops in the long bone *i.e.* tibia and femur (near the knee joints) or the humerus (near the shoulders) [1]. It generally arises in early adolescence as in this period, bone grows rapidly; the risk of developing tumour is more. Osteosarcoma is more prevalent in boys than girls and is accounting about 80% of the primary skeletal sarcomas among all the malignancies of bone [2]. The bone sarcomas typically metastasize into the lungs

---

* **Corresponding authors Asif Jafri & Md. Arshad:** Molecular Endocrinology Lab, Department of Zoology, University of Lucknow, Lucknow, India; E-mails: asifjafri.jafri@rediffmail.com, arshadm123@rediffmail.com

Sandeep Kumar & Dhiraj Kumar (Eds.)

and then metastasize to the lymph nodes, which are found in isolated cases [3].

Therefore, intensive research is required to develop a novel anti-cancerous drug from various biological or synthetic resources to heal this disease. The research on some phytochemicals has revealed that they possess potent anti-proliferative and apoptotic properties and will be helpful in the development of future drugs for cancer patients without any side effects. The methods described in this chapter deals the protocols of culture, handling and maintenance of osteosarcoma cells in the laboratory as well as to detect the apoptosis through the potent markers *viz.* DNA fragmentation and Caspase-3 activation.

Osteosarcoma Cells Culture and Maintenance

- At first, the animal cell culture laboratory environment and all the media, buffers and equipment are to be made sterile to prevent any microbial or fungal contamination in the culture of cells.
- The osteosarcoma cells, MG-63, SaOS$_2$ *etc.* are to be procured from any national or international cell repositories.
- The osteosarcoma cells are cultured in the desired culture medium, prepared with 89% culture media powder complemented with 1.5 g/l of sodium bicarbonate, 2mM of L-glutamine, 0.1mM of nonessential amino acids (NEAA), 1.0mM of sodium pyruvate, 1% antibiotic (penicillin-streptomycin) solution, and 10% (v/v) of heat-inactivated fetal bovine serum (FBS).
- The osteosarcoma cells are cultured at 37°C temperature, 5% CO_2 and 95% humidity in a tissue culture incubator.
- Cells are kept in the culture flask for proliferation and as it approaches up to 75-80% confluence level, then it is to be passaged or sub-cultured followed by splitting of some cells into a new cell culture flask supplemented with fresh culture media.
- This sub-culturing of cells will make available a continuous progression of osteosarcoma cells to perform various experiments of proliferation and apoptosis *via* the potent molecular markers.

DNA Fragmentation

Apoptosis is a type of dynamic and naturally controlled manner of cellular death that eliminates unwanted cells. In an apoptotic progression, the characteristic morphological alterations *viz.* cellular and nuclear shrinkage/rounding off cells, formation of apoptotic cell bodies and chromatin condensation/fragmentation [4]. The fragmentation of chromosomal DNA into smaller size oligo-nucleosomal fragments is a typical biochemical hallmark of programmed cell death [5]. In DNA fragmentation, the double-stranded DNA is cleaved into multiples of 180 base pair lengths by an endonuclease activity. A typical "ladder" pattern is observed when this apoptotic fragmented sample is electrophoresed on an agarose gel. DNA degradation of this type is usually found in cells that are morphologically apoptotic and the generation of "ladder" has been extensively used to define apoptosis in many experimental systems [6]. The apoptotic cells will form a typical DNA laddering pattern, whereas necrotic cells may produce a smear pattern. The DNA of the viable cells will stay compacted as it has a higher molecular weight band and thus lies on the upper part of the gel.

The major steps involved in DNA fragmentation analysis are as follows.

i. **Seeding of the Osteosarcoma Cells in 6 Wells Culture Plate:**
   ○ At first, the used culture media is to be removed from the confluent cell culture flask.
   ○ Gently wash the osteosarcoma cells with phosphate buffer saline (PBS) in such a way that it is not disrupting the layer of cells.
   ○ Then remove the wash solution and add the pre-warmed Trypsin EDTAsolution to the side of the flask.
   ○ The flask is kept for about 2 min at 37°C in the incubator and then observes the cells under the inverted microscope for detachment.
   ○ When the cells are in suspension, add 2-3 folds volume of complete culture media to neutralize the effect of Trypsin.
   ○ Then the detached cells are taken into a 15 ml falcon tube and centrifuge them for 5 min at 1000 rpm.
   ○ The supernatant is removed and the pellet is re-suspending into a 1 ml of complete culture medium and counts the cells by Haemocytometer.
   ○ Dilute the cell suspension and seeding the osteosarcoma cells at a density of 1×10^6 cells per well in a 6 wells culture plate and kept the plate into an incubator for the next day.

ii. **Extraction of Fragmented DNA:**
   ○ Induce apoptosis in the seeded osteosarcoma cells with the promising natural or synthesized compounds at the different concentration for 24 h in 5% CO_2 incubator and at 37°C humidified atmosphere.
   ○ After the treatment period, harvest the cells by trypsin EDTA and centrifuge them for 5 min at 1200 rpm at 4°C.
   ○ Discard the supernatant and re-suspend the pellet in 1 ml of ice-cold PBS (1X) and transfer it into a 1.5 ml micro centrifuge tube. Centrifuge the cells at 1200 rpm for 3 min at 4°C.
   ○ Aspirate the PBS and then lyse the cells with the help of 500 µl of DNA lysis buffer (20 mM EDTA, 10 mM Tris-HCl pH 8.0, 0.2% Triton X-100 and 100 µg/ml proteinase K) and keep the samples at 37°C for 1.5 h.
   ○ Centrifuge the samples at 10,000 rpm for about 10 min and collect the

supernatant containing DNA into a new microcentrifuge tube.
- Add an equivalent volume of isopropanol and 25µl 4M NaCl to a final concentration 100mM.
- Incubate the sample for overnight at -20°C and then centrifuge it at 10,000 rpm for 20 min at room temperature.
- Dissolve the pellet in 50 µl of double distilled water (ddH$_2$O) containing 2 µl of RNase A (10 mg/ml) and then re-incubated for 1 h at 37°C.

iii. **Agarose Gel Electrophoresis for Analysis of Fragmented DNA:**
- Add1.5 g of agarose in 100 ml of 1× TAE buffer and heat it until it gets fully dissolved. Add 0.5 µg/mlethidium bromide (EtBr) to this solution and then pour it into the gel-casting tray.
- After 20-30 min when the gel gets polymerised take out the gel comb and places the gel into an electrophoresis tank containing TAE buffer (1X) to cover the gel about 1 mm.
- Gently load all DNA samples with loading dye (Bromophenol blue or xylene cyanol) into each well, respectively and one well with 1 kb DNA ladder/marker.
- Run the gel which improves the resolution of DNA fragments.
- Attach the supply so that the DNA migrates towards the anode and electrophoreses at a low voltage *i.e.*, 50 V for approximately 2 h that improves the DNA fragments resolution on a gel.
- Switch off the power supply when the loading dye has migrated approximately two-thirds of the way down the gel.
- The gel is finally analysed and documented by photography under the UV trans-illuminator or ultraviolet illumination gel-doc System for DNA fragmentation.

Detection of Apoptosis *via* Caspase-3 Activity in Cancer Cells

The Caspase-3 or cysteine-aspartic acid protease activity is a typical biochemical marker in the apoptotic signalling. The caspase-3 is an important downstream effector caspase and plays a key role in the implementation of programmed cell death by cleaving the cellular substrates [7]. Caspase-3 is also known as CPP32, Yama or Apopain. Caspase-3 is associated with the cysteine proteases family and the member of interleukin 1β enzyme. The Caspase-3 is generally present in the cells as a pro-caspase-3, an inactive proenzyme of 32 kDa. During the apoptosis process,the up regulatory proteases *viz.* Caspase-6, Caspase-8 and Granzyme B fragment this pro-caspase-3into active 17 and 12 kDa subunits. Poly ADP ribose polymerase (PARP), nuclear lamin, sterol regulatory element binding proteins (SREBPs) etc are some othercaspase-3 downstream substrates. The Caspase-3 plays an important role in the apoptosis; as caspase-3 overexpression can induce the apoptosis whereas, its inhibition checks the cells from entering into an apoptotic pathway.

The caspase-3 protease colourimetric assay recognizes the specific amino acid sequence DEVD (Asp-Glue-Val-Asp) that are the caspase-3 upstream cleavage site in PARP [8]. The composition of DEVD-pNA substrate includes p-nitroani-lide(pNA), chromophore, and a synthetic tetra-peptideDEVD.The caspase-3/related caspases cleavages the substrate DEVD-pNA, the intensity of light absorbance of free pNA is quantified by using a microplate reader at 405 nm wavelength. The comparison between the absorbance of pNA from the apoptotic cells with control (untreated sample) provides convenient data for determining the fold increase in the caspase-3 activity.

The steps involved in caspase-3 apoptosis detection in osteosarcoma cells are as follows:

- At first, 3×10^6 osteosarcoma cells/ml are plated in 6 wells cell culture plate and induce the apoptosis with different concentrations of the test compound as we have previously discussed in DNA fragmentation assay.
- After treatment harvests the cells and washed twice with cold PBS.
- Now the cell pellets are re suspended in50 μlice-cold cell lysis buffer and keep it on the ice for about 10 min.
- Centrifuge the samples at 10,000 rpm for 2 min and then transfer the supernatant having cytosolic extract into a new microcentrifuge tube.
- The reaction is carried out in two types of wells *viz.* sample wells and background wells.
- Add 50 μl of each cell lysate (sample) are aliquoted into the sample wells of 96 wells cell culture plate and for the background, wells add 50 μl reaction buffer.
- Now add 50μl reaction buffer having 10 mM DTT to each well and incubate the plate on ice for further 30 min.
- Then 5 μl of 4 mM DEVD-pNA substrate is mixed into each well and incubate the sample at 37°C for about 2 h.
- After the incubation read the absorbance with the help of a microplate reader at the 405 nm wavelength.
- The caspase-3 activity can be evaluated by the comparison between the absorbance of treated samples with the untreated control group.

CONSENT FOR PUBLICATION

Not applicable.

CONFLICT OF INTEREST

The author confirms that this chapter contents have no conflict of interest.

ACKNOWLEDGEMENTS

Author Asif Jafri is thankful to Council of Scientific and Industrial Research, India for SRF Fellowship (File No. 09/107(0393)/2018-EMR-I) and Juhi Rais acknowledges Department of Science and Technology, India for DST INSPIRE Fellowship (IF140008). The authors are thankful to Central Instrumentation Facility of Department of Zoology, University of Lucknow, Lucknow, India.

REFERENCES

[1]   Mankin HJ, Gebhardt MC, Jennings LC, Springfield DS, Tomford WW. Long-term results of allograft replacement in the management of bone tumors. Clin Orthop Relat Res 1996; (324): 86-97.
[http://dx.doi.org/10.1097/00003086-199603000-00011] [PMID: 8595781]

[2]   Link MP, Goorin AM, Miser AW, *et al.* The effect of adjuvant chemotherapy on relapse-free survival in patients with osteosarcoma of the extremity. N Engl J Med 1986; 314(25): 1600-6.
[http://dx.doi.org/10.1056/NEJM198606193142502] [PMID: 3520317]

[3]   Nagarajan R, Kamruzzaman A, Ness KK, *et al.* Twenty years of follow-up of survivors of childhood osteosarcoma: a report from the Childhood Cancer Survivor Study. Cancer 2011; 117(3): 625-34.
[http://dx.doi.org/10.1002/cncr.25446] [PMID: 20922787]

[4]   Wong RS. Apoptosis in cancer: from pathogenesis to treatment. J Exp Clin Cancer Res 2011; 30: 87.
[http://dx.doi.org/10.1186/1756-9966-30-87] [PMID: 21943236]

[5]   Yeung MC, Printen JA. Accelerated apoptotic DNA laddering protocol. Biotechniques 2002; 33(4): 734-, 736.
[http://dx.doi.org/10.2144/02334bm03] [PMID: 12398177]

[6]   Martin SJ, Lennon SV, Bonham AM, Cotter TG. Induction of apoptosis (programmed cell death) in human leukemic HL-60 cells by inhibition of RNA or protein synthesis. J Immunol 1990; 145(6): 1859-67.
[PMID: 2167911]

[7]   Ahamad MS, Siddiqui S, Jafri A, Ahmad S, Afzal M, Arshad M. Induction of apoptosis and antiproliferative activity of naringenin in human epidermoid carcinoma cell through ROS generation and cell cycle arrest. PLoS One 2014; 9(10)e110003
[http://dx.doi.org/10.1371/journal.pone.0110003] [PMID: 25330158]

[8]   Jafri A, Siddiqui S, Rais J, *et al.* Induction of apoptosis by piperine in human cervical adenocarcinoma *via* ROS mediated mitochondrial pathway and caspase-3 activation. EXCLI J, 2019; 8: 154-64.
[PMID: 31217779]

CHAPTER 12

Culture and Maintenance of Human Ovarian Carcinoma Cells for Scrutinizing Anti-cancerous Activities of Various Compounds *via* Some Potent Molecular Markers

Juhi Rais[*], **Asif Jafri, Madhu Tripathi** and **Md Arshad**[*]

Molecular Endocrinology Lab, Department of Zoology, University of Lucknow, Lucknow, India

Abstract: Ovarian carcinoma is the 5th most common type of cancer of gynecologic origin and accounts for about one-fourth of the total malignancies of the female genital tract. Ovarian carcinoma accounts for highest mortality in females due to the development of chemo-resistance against drugs and lack of symptoms and undetectable biomarkers in the early stages of diagnosis. Tumour debulking, chemotherapies, radiotherapies, targeted therapies, immunotherapies and stem cell transplants are some of the measures that have been adopted by the experts for curing the disease but still, full control over the problem has not been achieved. Research on various herbal and chemosynthetic nano-compounds have shown a new light in this regard, as the studies on them so far have revealed that they have anti-proliferative and apoptotic properties that will help in finding new ways to develop drugs for cancer patients. This chapter deals how to culture and maintain the human ovarian carcinoma cell lines in the laboratory which are being procured from cell repositories and then to study the anti-cancer efficacy of various promising compounds by potent molecular markers like cell-cycle progression and annexin V- FITC apoptosis detection.

Keywords: Anti-proliferative properties, Annexin V-FITC, Nano-compound, Apoptosis, Anti-cancer efficacy, Cancer cell line, Cell-cycle progression, Molecular markers, Ovarian carcinoma.

INTRODUCTION

Ovarian carcinoma is the 5th most common type of cancer of gynecologic origin and accounts for about one-fourth of the total malignancies of the female genital tract. Chemo-resistance and lack of early prognostic markers account for the low survival rate. Less than 25% of ovarian carcinoma is diagnosed at stage I. It is usually diagnosed at an advanced stage when the cure rate is less than 20% [1].

[*] **Corresponding authors Juhi Rais & Md. Arshad:** Molecular Endocrinology Lab, Department of Zoology, University of Lucknow, Lucknow, India; E-mails: juhirais44@gmail.com and arshadm123@rediffmail.com

Sandeep Kumar & Dhiraj Kumar (Eds.)

Different clinical expression, pathological characteristics and genetic aetiology signify that an ovarian carcinoma is a heterogeneous group of malignancy [2, 3].

Endeavours to improve survival continue to focus research on various herbal and chemosynthetic nano-compounds which have shown a new light in this regard. The studies on them so far have revealed that they have anti-proliferative and apoptotic properties that will help in finding new ways to develop drugs for cancer patients. The following chapter presents a protocol that deals how to culture and maintain the human ovarian carcinoma cell lines in the laboratory and to study the anti-cancer efficacy of various promising compounds by potent molecular markers like cell-cycle progression and annexin V- FITC apoptosis detection.

I). *In Vitro* Culture and Maintenance of Ovarian Carcinoma Cells

- Before procuring the cells, the cell culture laboratory's environment is to be made aseptic. All media, supplements and reagents must be sterile to prevent microbial growth in the cell culture.
- The Human ovarian carcinoma cell line like OVCAR-3, PA-1, Caov-3, SW626, SK-OV-3 *etc.* are to be obtained from the different cell repositories.
- Before starting the work one should check the information that is given with the cell line to identify what media type and additives are to be used.
- The cells are cultured in 90% media powder supplemented with 2 mM of L-glutamine, 1.5 g/l of sodium bicarbonate, 0.1 mM of nonessential amino acids, 1.0 mM of sodium pyruvate, and 10% of heat-inactivated FBS. The culture media is supplemented with 1% antibiotics solution.
- The ovarian carcinoma cells are incubated at 37°C in a 95% air/5% CO_2 and water-saturated atmosphere.
- Cells are kept in the incubator until they get proliferated and occupy all of the available substrate (until the flask gets 80-90% confluent).
- Once the flask gets confluent, the cells have to be sub-cultured (*i.e.*, passaged) by transferring them to a new culture flask with fresh growth medium.
- The sub-culturing of cells will provide continued growth to the cell line and hence different desired experiments can be performed.

II). Cell Cycle Analysis

One of the earliest applications of flow cytometry is the cell cycle analysis. A variety of DNA binding dyes can stain the DNA of mammalian cells. DNA-binding dyes include propidium iodide (PI), 7-aminoactinomycin-D (7-AAD), Hoechst 33342 and 33258, TO-PRO-3, 4'6'-diamidino-2-phenylindole (DAPI), *etc.* These dyes bind in proportion to the amount of DNA present in the cell. So the cells that are in S phase will have more DNA than cells in G_1 and G_2 has approximately twice cells as that in G_1. The cells will thus proportionally take up more dye and will fluoresce more brightly until they have doubled their DNA content. To allow the entry of the dye cells must be fixed or permeabilized otherwise it is actively pumped out by living cells.

Steps involved in cell cycle analysis using DNA binding dye (PI).

a. Seeding of the Ovarian Carcinoma Cells in 6 wells plate:
   ○ The very first step is to remove and discard the spent cell culture media from the culture flask.
   ○ Gently wash the cells using PBS such that it won't disturb the cell layer of the flask.
   ○ Then remove and discard the wash solution from the culture flask.
   ○ Add the pre-warmed dissociation reagent such as trypsin EDTA to the culture flask and incubate it at the room temperature for approximately 2 min.
   ○ Observe the cells under the microscope for detachment. If cells are less than 90% detached, increase the incubation time a few more min, checking for dissociation every 30 secs.
   ○ Add twice the volume of pre-warmed complete growth medium.
   ○ Take a 15 ml conical tube and transfer the cells into it and then centrifuge them at 1000 rpm for 5 min.
   ○ Re-suspend the cell pellet in a minimal volume of pre-warmed complete growth medium and count the cells using a Haemocytometer.
   ○ Dilute cell suspension to the seeding density of 1×10^6 cells in 6 wells culture plate and keep the suspended cells into the incubator.
b. Analysis of cellular DNA contents by flow cytometry:
   ○ Treat the seeded cells with desired organic/herbal compound and incubate it for 24 h in 5% CO_2 incubator and at 37°C temperature.
   ○ After 24 h of incubation, harvest the cultured cells and centrifuge them at 1200 rpm for 5 min.
   ○ Re-suspend the pellet in 600 μl of cold PBS (1X) and then add 70% of chilled Ethanol so as to fix the cells and keep it for 2 h.
   ○ Centrifuge the cells at 1200 rpm for 4 min.
   ○ Discard the supernatant and then add 400 μl of lysis buffer (1X PBS + 0.2% Triton X-100).
   ○ Incubate it for 30 min at 4°C.
   ○ Centrifuge it at 1200 rpm for about 4 min.
   ○ Discard the supernatant and re-suspend the pellet in 1ml of 1X PBS containing 20 μl of RNase (10mg/ml)
   ○ Incubate the cells for 30 min at 37°C and then centrifuge it at 1200 rpm for 10 min.

○ Re-suspend the pellet in 500µl of 1X PBS containing 10µl of DNA binding dye (PI) followed by incubation for 30 min at room temperature in dark.
○ Measure the PI fluorescence of individual nuclei using flow cytometer and quantify the results by the software that is inbuilt in the flow cytometer.

III). Detection of Apoptosis in Cell Culture System *via* Annexin V- FITC

A normal physiologic process by which cells commit suicide by activating its intracellular death program is termed as apoptosis or programmed cell death. This process occurs during embryonic development as well as in the maintenance of tissue homeostasis. The apoptotic program is different from necrosis by loss of plasma membrane symmetry, shrinkage of the cytoplasm and nucleus, and inter-nucleosomal cleavage of DNA. The early stage of apoptosis is the loss of phospholipid symmetry. In normal live cells, the inner leaflet or the cytoplasmic surface of the cell membrane contains phosphatidylserine [4]. However, this phosphatidylserine exposes itself to the external portion of the membrane as it translocates from the inner leaflet of the plasma membrane to its outer membrane. A group of homologous proteins called the annexins, which in the presence of calcium, binds with the phospholipids. Annexin V-FITC is a fluorescent probe which in the presence of calcium binds to this phosphatidylserine [5].

The procedure starts as Annexin V-FITC binds to the phosphatidylserine of the outer membrane of cells that begins the apoptotic process whereas, in the cells where the cell membrane has been totally compromised, the cellular DNA binds the propidium iodide. The cells are then incubated with Annexin V-FITC and propidium iodide. Analysis of Flow cytometry reveals that Annexin V-FITC is detected as a green fluorescence which shows the apoptotic cells and propidium iodide is detected as a red and green fluorescence which indicates the dead cells and live cells show little or no fluorescence [5].

The protocol to detect apoptosis in cell culture system *via* Annexin V- FITC is as follows:

• The cells are to be seeded and treated with the test compound as before we did for cell cycle analysis.
• After the treatment, harvest the cells and wash it with cold phosphate-buffered saline (PBS).
• Then prepare 1X Annexin-binding buffer (for 10 assays, 1 ml 5X Annexin binding buffer is added to 4 ml de-ionized water).
• From 5 µL of 1 mg/ml PI stock solution, 100 µg/ml working solution of PI is prepared by diluting it in 45 µL 1X Annexin-binding buffer.
• Centrifuge the washed cells at 1200 rpm for 5 min, discard the supernatant, and re-suspend the cells by determining the cell density through haemocytometer

and dilute in 1X Annexin-binding buffer to ~1 × 10⁶ cells/ml so as to prepare a sufficient volume to have 100 µL per assay.

- Now, add 5 µl of Annexin V- FITC and 1 µl of the 100 µg/ml PI working solution to each 100 µl of cell suspension.
- Incubate the cells for 15 min at room temperature and then add 400 µl of 1X Annexin-binding buffer and mix gently.
- Keep the samples on ice and analyze the stained cells as soon as possible by flow cytometry.
- The flow cytometer will separate the cells into three groups:
  - live cells will show only a low level of fluorescence.
  - apoptotic cells bind with Annexin V-FITC, shows green fluorescence.
  - dead cells bind with propidium iodide shows both red and green fluorescence.

CONSENT FOR PUBLICATION

Not applicable.

CONFLICT OF INTEREST

The author confirms that this chapter contents have no conflict of interest.

ACKNOWLEDGEMENTS

The author Juhi Rais is thankful to Department of Science and Technology, India for DST INSPIRE Fellowship (IF140008) and author Asif Jafri acknowledges Council of Scientific and Industrial Research, India for SRF Fellowship (File No. 09/107(0393)/2018-EMR-I). The authors are thankful to Central Instrumentation Facility of Department of Zoology, University of Lucknow, Lucknow, India.

REFERENCES

[1]   Bast RC Jr. Status of tumor markers in ovarian cancer screening. J Clin Oncol 2003; 21(10) (Suppl.): 200s-5s.
[http://dx.doi.org/10.1200/JCO.2003.01.068] [PMID: 12743135]

[2]   Bast RC Jr, Hennessy B, Mills GB. The biology of ovarian cancer: new opportunities for translation. Nat Rev Cancer 2009; 9(6): 415-28.
[http://dx.doi.org/10.1038/nrc2644] [PMID: 19461667]

[3]   Cho KR, Shih IeM. Ovarian cancer. Annu Rev Pathol 2009; 4: 287-313.
[http://dx.doi.org/10.1146/annurev.pathol.4.110807.092246] [PMID: 18842102]

[4]   Koopman G, Reutelingsperger CP, Kuijten GA, Keehnen RM, Pals ST, van Oers MH. Annexin V for flow cytometric detection of phosphatidylserine expression on B cells undergoing apoptosis. Blood 1994; 84(5): 1415-20.
[http://dx.doi.org/10.1182/blood.V84.5.1415.1415] [PMID: 8068938]

[5]   Vermes I, Haanen C, Steffens-Nakken H, Reutelingsperger C. A novel assay for apoptosis. Flow cytometric detection of phosphatidylserine expression on early apoptotic cells using fluorescein labelled Annexin V. J Immunol Methods 1995; 184(1): 39-51.
[http://dx.doi.org/10.1016/0022-1759(95)00072-I] [PMID: 7622868]

CHAPTER 13

Dictyostelium Discoideum: Live Cell Imaging in Changing Perspective

Abhishek Singh[*]

School of Life Sciences, Jawaharlal Nehru University, New Delhi, India

Abstract: The advent of advanced microscopes; during microscope evolution from simple microscopes to confocal and live cell microscope; having digital imaging facility revolutionized our view for the living cells. In the protein localization study, fluorescent proteins are tagged at amino or carboxyl (preferably) terminal of desired protein for live cell study. These live cell studies improved our understanding of protein dynamics and understanding its role in biological regulation. The mutational variants of fluorescent tags (GFP, RFP); can be used with different protein; which will efficiently use UV-Visible to Far Red light spectrum; without overlapping of excitation and emission spectrum. Further, various cell organelle (Lysosome, Golgi bodies, Endoplasmic Reticulum, Mitochondria, Nucleus) trackers; improved our live cell localization studies in the wide non-overlapping UV-Visible spectrum.This chapter gives an overview for live cell protein localization study in mitotically active, unicellular stage of *Dictyostelium discoideum*. This evolutionary cutting edge organism had both unicellular as well as multicellular stages during its life cycle. This chapter will provide the design of fusion of fluorescent tag to the specific gene and its live cell localization. Further, it will cover; transformation of the unicellular organism; drug based selection; sample preparation with nuclear, mitochondrial localization markers (trackers) and live cell localization study on live cell-confocal microscope setup. It will also have a glimpse of the design of fusion protein with an aspect of advantage and disadvantages.

Keywords: Confocal, DAPI, *Dictyostelium discoideum*, EMCCD, Fluorescent proteins GFP, Golgo bodies, RFP, Live cell imaging, MitoTracker, Mitochondria, UV-visible spectrum.

AN INTRODUCTION TO *DICTYOSTELIUM DISCOIDEUM*

Kenneth Raper, first isolated *Dictyostelium discoideum* from the deciduous forest soil and decaying leaves from North Carolina, the USA in the year 1935. In its natural habitat, amoebic *Dictyostelium* cells feed on bacteria and yeast by chemotaxis tracking and multiply by mitosis. It provides a valuable model system

---

[*] **Corresponding author Abhishek Singh:** School of Life Sciences, Jawaharlal Nehru University, New Delhi, India; E-mail: abhishekchauhan101@gmail.com

Sandeep Kumar & Dhiraj Kumar (Eds.)

for studying development, differentiation, chemotaxis, phagocytosis and signal transduction. The genome of *Dictyostelium discoideum* has been sequenced and shows the presence of nearly 12,500 proteins coding genes [1].

The Life Cycle of *D. Discoideum*

D. discoideum has a very short life cycle as a facultative multicellular organism, which has vegetative growth until the food is available (Fig. **1**) and enters multicellular development as soon as nutrient gets depleted. The life cycle of amoebae exists in the unicellular vegetative cycle, multicellular social developmental cycle (aggregation, mound, slug, Mexican hat, early culminant and culminant stages) and sexual cycle stages.

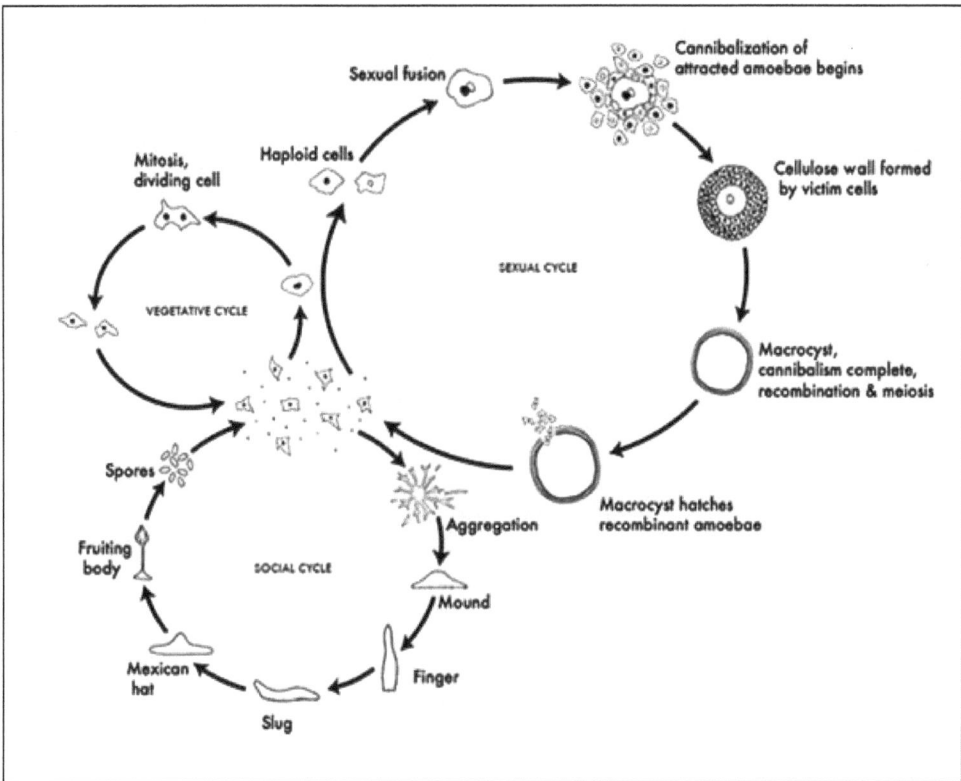

Fig. (1). *Dictyostelium discoideum* **life cycle has three possible phases**-Vegetative, asexual (social) and sexual. The abundance of food makes amoebae to reside in single cell vegetative cycle. But, the sexual or social (asexual) developmental cycle decided by the availability of appropriate mating types is largely triggered by starvation. This developmental cycle is used by the amoebae as a protection strategy during unfavourable conditions and after completion of a cycle, an abundance of nutrients survivors to continue vegetative phase (Brown D, Strassmann JE. CC3.0 copyright) (http://www.dictybase.org/Multimedia/ DdLifeCycles/index.html [2].

Advantages of *Dictyostelium Discoideum* as a Model System

There are several reasons for using *Dictyostelium discoideum* as a model system.

- It occupies a strategic phylogenetic position as it has both unicellular and multicellular stages incorporated in its life cycle. Thus, this organism provides a link between unicellular to multicellular transitions. It is suitable for the study of pattern formation and morphogenesis in the multicellular entity.
- It is being used as a model system to study chemotaxis and phagocytosis as it is a eukaryotic system and shows similarity to macrophages in these aspects.
- Unicellular amoebic cells are easy to culture in an inexpensive HL5 media. It has been extensively used as a model system to study developmental aspects and cellular processes.
- It is experimentally accessible and very permissive for genetic manipulations as they are haploid. Haploid genome makes it easy to generate and select mutants of function associated genes. One could apply reverse genetics methods to establish a direct link between functions and genes.
- The genome of 34 Mb size is divided into six haploid set of chromosomes, is fully sequenced and has small introns and small non- coding intergenic regions. The whole genome sequences (protein, genomic DNA, cDNA) are available free online resource at- www.dictybase.org, making it a very tractable system.
- The genome is highly AT-rich (77.57%). Coding region possesses ~40% GC content while the non-coding region possesses only ~10% GC content. It is one of the high gene dense (~1 gene per 2.6 kb) genomes. The introns are usually small (100-200bp) and few in numbers.
- Temporal separations of the vegetative and developmental phases allow many lethal developmental mutants to be propagated under vegetative conditions thus behaving like conditional mutants [3].
- It is a good model system for biomedical research. At least, 33 *D. discoideum* genes have been identified as orthologs of disease-related genes which can be studied in this model organism [1, 4].
- Itis being used for screening of genes responsible for drug resistance. Five genes have been identified in *D. discoideum* which confers cisplatin (an anticancer drug) resistance [5].
- Many human pathogens can use *D. discoideum* as a host so it can be used for the study of host-pathogen interactions [6].

The most common wild type strains Ax2 and Ax3 of *D. discoideum* amoebae render growth simply in the laboratory either in HL5 growth medium or with the lawn of bacteria. The amoebae grow optimally at 22°C and handled aseptically with temperature-controlled culture room or incubators. Amoebae with cryopreservatives glycerol and dimethyl sulfoxide (DMSO)under liquid nitrogen and spores with silica gel at 4°C can be stored easily for a long time [7].

The Procedure of *Dictyostelium discoideum Ax2* Cells Culture: (Adapted [8]):

1. The temperature controlled culture room or incubator (22°C) is a primary requirement for the growth and development of *Dictyostelium discoideum* and all media should be at similar or at room temperature for ready to use.
2. Aseptically grow *Dictyostelium* axenic Ax2 strain cells at 22°C in HL5 medium either in 90mm tissue culture dishes or in shaken suspension with about 50% confluency.
3. Transfer cells to a 250ml tissue culture flask having 100ml HL5 medium using sterile pipet.
4. Good aeration for growth of cells can be provided by ~2.5 ratio of flask/medium volume.
5. Incubate at 22°C under shaking at 200 rpm until log phase is reached.
6. Under shaking condition cells with a doubling time of 10 to 12 hr reach in log-phase in ~2 days with a density of 5×10^6 cells/ml.
7. Centrifuge cells at 22°C with $1500 \times g$ for 4 min to collect log-phase cells.
8. Passaging of mass culture plates required for every 3 to 5 days with a split ratio of ~1:5 in the HL5 medium.

Required Materials

- *Dictyostelium Discoideum* Ax2 Strain Cells (Dictybase)
- HL5 medium
- KK2 buffer
- 90-mm tissue culture plate
- 250-ml tissue culture flask
- 22°C Incubator and Shaker

Growing of *Dictyostelium Discoideum*on a Bacterial Lawn

1. *Dictyostelium discoideum* can also be grown on bacterial lawns to study phagocytosis [7]. Procedure for growing amoebic cells on the bacterial lawn as described below:
2. Grow *Klebsiella aerogenes* bacterium cells from a single colony for over night at room temperature or at 37°C in LB medium without dihydrostreptomycin with shaking at 250 rpm.
3. Around 10^6 log-phase *Dictyostelium discoideum* amoebae mixed with 200 μl of above grown *Klebsiella aerogenes* bacterial culture and spread them on a 90-mm SM agar plate.

4. The procedure can be scaled up as needed by increasing the number of plates used.

5. Incubate these plates in a humid environment at 22°C for 36 to 48 h. recover $\sim 10^9$ growing amoebic cells before clearing of bacterial lawn and entering into the development programme from 90 mm plate. Placing the plates in a closed plastic box in the incubator will provide an appropriately humid environment. Within 36 to 48 h the bacterial lawn will be cleared by the growing cells and $\sim 10^9$ amoebae can be recovered per 90 mm plate. It is important to recover the amoebae before they eat up the bacteria and enter the development program.

6. Scrape amoebic cells from a plate, with ~5 ml of KK2 buffer to and transfer into a 50-ml conical tube and repeat this step 3-4 times.

7. Pellet down amoebic cells for 1 min at 1000×g and at 22°C. Discard the remaining turbid supernatant.

8. Re-suspend amoebic cells with 50 ml KK2 buffer and repeat steps 5 to 6 three to four times (until the supernatant is clear).

9. Use cleared pellet amoebic cell for subsequent study.

Required Materials

* *Klebsiella aerogenes* (ATCC)
* HL5 medium
* *Dictyostelium discoideum* Ax2 strain cells (Dictybase)
* 90-mm SM agar plates
* KK2 buffer
* Incubator 22°C

Recombinant Plasmid Construction

Dictyostelium discoideum amoebic transformed cell lines can be formed by either integrating gene of interest in the genome or by using episomal vectors [9, 10]. In the construction of episomal vectors the Ddp1 sequence (having amplification capacity of 100 copies per cell) and the cassette having actin 15 promoter/$2H_3$ terminator expression system used for the establishment of GFP fusion protein expressing cell lines [11 - 13].

In the protein distributional study of *Dictyostelium discoideum,* homologous recombination-mediated deletion or gene replacement gives a phenotype for the better access of the functionality of GFP fusion proteins

1. The constitutive episomal expression of the *Dictyostelium discoideum* gene can be done by following steps:

2. Identify the gene of interest by *in-silico* study using an online

platform/database (http://dictybase.org/).

3. Design forward and reverse primers for the open reading frame (ORF) region of the gene and synthesized by a company such as Sigma Aldrich.

4. Amplify full-length ORF from Ax2 genomic DNA by Polymerase Chain Reaction (PCR).

5. Restriction digestion of the purified PCR product as well as actin 15 promoter, Enhanced Yellow Fluorescent Protein (*Eyfp*) reporter and Geneticin (G418) selection gene containing *pB17S* vector at 37°C with BamHI and XhoI enzymes (Fig. **2**).

6. Ligate mixture of purified restriction digested PCR product of ORF and *pB17S* vector in the ratio of 3:1 respectively, by T4 DNA ligase and incubate at 22°C for 16 h.

7. Transform 100µL of competent cells of *Escherichia coli (E. coli)* DH5α strain; prepared by TSS method [14] with 5-10µLof ligation mixture and mixed by gentle tapping.

8. Incubate transformation mixture on ice for 20 min and followed with a heat shock at 42°C for 90 secs.

9. After heat shock cells were kept immediately on ice for 5 min and further incubated at 37°C for 1 h under shaking conditions at 200 rpm.

10. Plate transformation mixture on LB agar plates having appropriate selection pressure (Ampicillin, Kanamycin).

11. Incubate plates for overnight at 37°C.

12. Analyse positive transformants by different combinations of restriction digestions for identification and directional match of the present insert (ORF) in the vector and further confirmed by DNA sequencing.

13. Prepare DNA of positive construct at a mini scale for the transformation of the *Dictyostelium discoideum* Ax2 strain cells by electroporation.

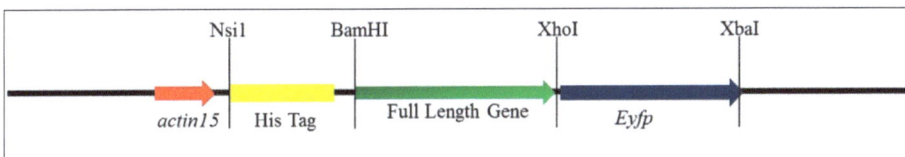

Fig. (2). Strategy for the cloning of the desired gene from *Dictyostelium discoideum* in the *pB17S* vector. **(A)** Schematic representation of cloning. The reporter *Eyfp* is present in between XhoI and XbaI restriction sites and amplified gene was ligated using BamHI and XhoI restriction sites upstream of it. This arrangement yields an *actin15* driven constitutive expression of desired with *Eyfp* reporter in the frame at the C-terminus. In this strategy, the 6xHis Tag is also inserted at the N-terminal.

Transformation of *Dictyostelium Discoideum* Cells by Electroporation (Based on [15])

1. It is very important to keep everything cold. Sterile H-50 buffer, cuvettes and sterile tubes containing the DNA should be prechilled on ice.

Note: The H-50 buffer may be substituted by E buffer.

2. Harvest fresh log phase cells of *Dictyostelium discoideum* Ax2 strain ($3\text{-}5\text{x}10^6$ cells/ml) grown in HL5 medium in the 90mm petri plate or suspension culture at 22°C after determining the number of cells/ml using a haemocytometer or an automated cell counter.

3. Wash gently ~$5\text{x}10^6$ cells twice with sterile chilled H50 buffer by centrifuging at 500xg for 3-5 min at 4°C. Resuspend pellet with a sterile pipette and centrifuge at 300xg for 2-3 min then resuspend in 100 µl of chilled H50 buffer having 4x107 cells.

4. Mix the chilled cell suspension with ~10µg construct plasmid DNA having less than 10% of total volume and pour into the 1mm size chilled electroporation cuvette (Sigma Aldrich USA).
   Note: Remember to transform a sample lacking DNA (negative control) as well as an empty vector positive control. A vector containing GFP alone should be used as a control.

5. Transfer 0.1 ml of the cell suspension to a pre-chilled sterile electroporation cuvette (1 mm gap width) and add 10µg of supercoiled plasmid DNA preferably in 7-12µl of TE buffer and then mix by gently pipetting up and down avoiding introducing air bubbles to the cell suspension. Return the cuvette to ice.

6. Set the Bio-Rad Gene Pulser™ electroporator at 0.65-0.70kV/25µF capacitance with infinity (∞) resistance and electroporate cells twice by giving pulse with the time constant of 0.5ms with a time interval of 15 secs. Electroporation of cell/DNA mixture should be done quickly to reduce sedimentation of cells in the cuvette.
   Note: Optimization of the electroporator devices may be needed according to the manufacturer guidelines. Before giving the electric pulse; removal of excess moisture from electrodes (cuvette) with the help of tissue is needed.

7. After sec pulse incubates cells immediately on ice for 5 min.

8. Add 800µl fresh HL5 medium into the cuvette having electroporated cells, mix them by pipetting up and down and withdraw HL5 medium having cells from the cuvette and add in to the petri plate having ~12ml HL5 media. Repeat once more similarly for surety of collected cells. A gentle swirling of dish evenly distributes the cells. Incubate cells in petri plates for 24 h at 22°C for the attachment and to come out from shock.
   Note: Cell death might be expected upon electroporation. However, the cell survival rate should be above 25%.

9. Apply appropriate antibiotic (Geneticin-G418) selection pressure after 24 h of incubation of cells at 22°C initially at 5µg/ml concentration and remove dead cells after 2-3 days by changing media through aspiration.

10. Gradually increase the concentration of antibiotic (Geneticin-G418) up to

maximum 200µg/ml (depending on experiment requirement and the response of higher level of expressed protein in the cells) and remove dead cells after 2-3 days by changing media through aspiration.

11. Isolate clones from the confluent plate by clonally plating around 100 transformed amoebae cells on *Klebsiella aerogenes* lawn in 90mm SM agar plate.

12. Grow amoebae to make clones for 3-4 days at 22°C in a humid environment in the incubator.

13. Transfer individual clones by scrapping feeding edge of plaques into 24 well plates having 20µg/ml Geneticin-G418 in HL5 and allow cells to grow in HL5 at 22°C.

14. Analyse clones by a screening of optimum expression in transformants by taking cells out and allow them to attach for 15 min in chambered coverslip and examine under confocal live cell imaging microscope for the present as well as the distribution of the GFP fusion protein.

15. Permanent stock made by storing spores at -80°C in freezing storage solution after development on NNA containing 10µg/ml Geneticin-G418.

Materials for the Transformation of *Dictyostelium Discoideum*

Dictyostelium discoideum cells grown to log phase in HL5 liquid medium, Ice cold H50 buffer and Electroporation buffer, Supercoiled DNA vector encoding a GFP fusion protein, 90mm petri plate, 90mm NNA plate, Healing solution, HL5 medium, 20 mg/ml (w/v) G418 solution, Electroporation cuvettes (0.2-cm electrode gap; *e.g.*, Bio-Rad), Electroporator (*e.g.*, Gene Pulser II, Bio-Rad), 22°C Incubator with shaker, 24-well plates, Haemocytometer or Automated Cell Counter for counting cells, Microscopic visualization in both bright field and fluorescence, *Klebsiella aerogenes* prepared for cloning of positive transformants as well as for the revival of cells.

NOTE: All manipulations are performed aseptically on ice.

Live Cell Imaging Microscopy

The green fluorescent protein (GFP) discovered from *Aequorea victoria*, had revolutionized cell biology by live cell imaging studies through labelling cells with recombinant fusion GFP and other variant proteins [16 - 19]. This fluorescent protein, as well as chemically synthesized tag labelling, played as a prime driving force for innovation in the fluorescence light microscopy.

In the wide-field epifluorescence microscope; the problem of out-of-focus light and high fluorescent cytoplasmic background has risen the demand for inventing confocal microscopes with the ability of rapid multidimensional imaging of living cells. The pinholes are placed in the emission light path in the confocal microscope largely allows light to pass which originates from point of focus.

One of the variants of a confocal microscope is confocal laser scanning microscope (CLSM); that is based on single excitation laser beam rapid scanning of the specimen and quantification by a photomultiplier tube (PMT) followed with image reconstruction by a computer [20]. Another variant of a confocal microscope is spinning disk confocal microscope (SDCM) has rapidly rotating spinning disk named as Nipkow disk with thousands of pinholes as points of a simultaneous light scan of specimen and quantification by charge couple device (CCD) camera. However, due to the problem of inefficient transmission of the excitation light through pinhole disk to the specimen with the adequate bright confocal signal; the spinning disk systems were not a good choice for the fluorescence microscopy. This problem was resolved by introducing Yokogawa dual spinning disk design in combination with high-intensity laser as light sources and highly sensitive digital cameras like charge couple device (CCD) cameras. In the Yokogawa dual spinning disk design, the arrangement of the pinholes is in a spiral array of the equal pitch; that gives uniform illumination to the specimen. In this, disk design, the incoming excitation light is focused through pinholes by a sec disk having microlenses [21].

An SDCM system has several attributes over CLSM system for fluorescence-based live cell imaging. One reason is faster parallel scanning of thousands of point of the specimen with higher frame acquisition rates; theoretically as possible as 2000 frames per sec (FPS) by highly sensitive cooled charge-coupled device (CCD) cameras compared to slower scanning and less sensitive detection by PMT in CLSM. The PMT has limitations of low quantum efficiency and high noise due to charge multiplication; even it can achieve extremely high gain. While cooled CCD cameras produce better and low noisy images. Even development of high resolution and low noise scientific grade metal oxide semiconductor cameras are promising for the future.

SDCM is less prone to fluorescence saturation in which additional excitation light contributes to photo damage and not yields into the additional signal. In the typical camera exposure for a few hundred milliseconds in the SDCM system; the image formed by the results of illumination of several hundred times with short exposures by low peak illumination intensity. Contrastingly, in the CLSM higher fluorescence saturation occurs due to the illumination of each point once in the acquisition per frame leading to the elongated stay time per pixel with very high fold illumination intensities and it is likely the reason more photo bleaching compared with SDCM [22].

The disadvantage of SDCM over CLSM is the reduction in the confocality due to the transmission of the out-of-focus emission light to the camera *via* adjacent pinholes. Additionally, in the SDCM system, larger pinhole size for lower magnification due to their optimization for higher (100x) magnification causes transmission of the out-of-focus emission light to the camera which further reduces confocality. But this trivial loss of confocality is compensated by reduced photo bleaching and enhanced signal to noise ratio [23].

Live Cell Imaging Microscopy of *Dictyostelium Discoideum* Amoebic Cells (Adapted from [8])

- Live cell imaging is an outcome of detection of the fluorescent fusion protein as a signal to noise ratio and damage to the cells caused by free radicals produced by laser illumination due to interaction with cellular constituents or fluorescent tag/probe. The 10x and 20x magnifications have very little spatial resolution within the *Dictyostelium discoideum* amoebic cells due to approximately 10-15µM diameter but free radicals are not the problem.
- While 63x or 100x magnifications are desired for resolving subcellular localization of the tagged fluorescent protein or fluorescent probe within the *Dictyostelium discoideum* amoebic cells [8].
- High energy (short wavelength) lasers in the confocal microscopes, such as 405nm, 488nm lasers causes' damage by photobleaching and phototoxicity to single amoebic vegetative cells due to the excessive power of it. This problem can be reduced by keeping laser at minimum strength (power) like 2-10%; for shorter wavelength (high energy). While Long wavelength (low energy) laser-like 561nm can be used at higher power as much as 15% or more and by the opening of pinhole the signal can be increased. The developed cells are generally less sensitive to the photo-damage compared to unicellular vegetative amoebic cells caused by lasers [8].
- The free radicals can be scavenged or reduced by overnight incubating before imaging with 50-100µm ascorbic acid and 1 mM Trolox C (vitamin E analogue) containing growth medium. Moreover, the problem of photobleaching can be reduced by using an antifade mounting medium for the fixed cells and ICC and better images can be taken by slower scanning and averaging of better images [8].
- Even after taking above precautionary measures; if phototoxicity remains the problem then switching to spinning disc confocal microscope (SDCM) will be a better choice. In SDCM, the illumination *via* several small confocal volumes in the image acquisition of the whole sample decreases exposure leading to the reduced photobleaching and phototoxicity. Amoebic cells live movement can be followed at the fastest realistic frame with Andor iXon3 EMCCD camera as 1-3 frames per sec (FPS) and it can be adjusted according to the requirement of the

aim of the study like every frame of live cell movement capture at the interval of the 2-30 sec. The better images can be taken through slow scan by reducing scan area *via* decreasing default pixel area from 512 pixels x 512 pixels to 512 pixels x 300 pixels; leading to maximised frame rate with slower scans [24, 25]. The faster resolution with SDCM can be achieved by setting 488 nm laser power at 5%, the opening of pinhole up to 2.5 Airy units and collection of frames at 512 pixels x 300 pixels of promoter actin15-GFP fusion construct expression.

Instrument Design of Microscope, Stage and Control of the Environment

The Spinning Disk Confocal Microscope (SDCM) consists of a confocal head attached with Nikon Eclipse Ti inverted microscope for the observation of live cells in 35mm glass base petri plates. The EYFP fusion protein expressing in the *Dictyostelium discoideum* amoebae requires only 22°C temperature control unit; not the CO_2; that can be achieved by controlling room temperature or by controlling the temperature of the microscope chamber.

Note: Fluctuation in the temperature causes focus drift which affects higher magnification imaging experiments in the live cell imaging; for this reason, Nikon Eclipse Ti inverted microscope has Perfect Focus System (PFS) technology.

Spinning Disk Confocal Scanner and Illumination

The Yokogawa CSU-X1 spinning disk confocal scan head with higher adjustable disk rotation speed at very cool (-85 to -90°C) temperature with short camera exposure time; Andor iXon-3 Electron Multiply Charge Couple Device (EM-CCD)camera on Nikon Eclipse Ti microscope system.

Caution: Even, lower power laser-like 100mW can cause damage to the retina which requires appropriate safety precautions.

Note: In the live cell imaging experiments; minimum exposure time is most often dictated by the fluorescence signal that can be obtained from cells expressing Fluorescence-tagged proteins at physiological levels.

Photobleaching is the main problem of fluorescence live cell imaging which can be reduced by receiving maximum signal by selecting optimal emission filters. It can be minimized by reducing specimen exposure to excitation light by controlling laser shutters directly by the camera instead of through software. While the improved communication between Andor iXon3 camera and its IQ3 software with NIS Elements Eclipse Ti platform has a much shorter delay in the shutter opening and camera exposure in the time-lapse live cell imaging.

Image acquisition: In the compensation for low signal due to the inefficient transmission of excitation light in the Yokogawa spinning disk design and insufficient excitation because of laser power; the low light electron multiply charge coupled device (EMCCD) cameras are a good choice for live cell imaging. Cooling around -85 to -90°C of CCD chip in the Andor EMCCD camera minimizes dark current that leads to the achievement as the lowest read noise. Pinholes of 50 μm diameter give the appearance of the image of ~500nm diameter with 100x magnification after optimally focusing distance between the camera and spinning disk head with the Yokogawa disk.

Imaging of GFP-labeled Fusion Proteins in Live Single Cells

After checking the expression of EYFP fusion protein in the *Dictyostelium discoideum* cells; study of its cellular localization as well localization of nucleus with DAPI stain and mitochondria by MitoRed performed within single cells. The method of single cell lives cell imaging has been divided into two sections. The first part deals with a brief description of the imaging arrangement; while sec part has a detailed step-by-step description for sample preparation for live cell microscopy. In this live cell microscopy, the main emphasis is given to avoid the use of buffers like KK2 buffer, PBS buffer, due to the sensitivity of transformed amoebae cells having increased level of STRAP-EYFP fusion protein expressing; to the change in the osmolarity. In this live cell imaging experiment, the conditions were standardized for EYFP fusion protein over-expressing amoebae cells with their normal growth HL5 medium.

Materials

1. Transformed *Dictyostelium discoideum* cells over-expressing EYFP fusion protein grown to log phase in HL5 medium, Fresh HL5 Medium, KK2 buffer, 35mm glass base petri plate, Nuclear stain DAPI, Mitochondrial stain MitoRed, Haemocytometer for amoebae cell counting, Nikon Eclipse Ti Inverted microscope equipped with Electron Multiply charge-coupled device (EMCCD) camera, mercury light source, and appropriate filter for DAPI, FITC and TRITC wavelength, Computer with Andor IQ3 image acquisition software, standard mechanical stage, 22°C temperature regulator.
2. **Set up of imaging hardware:** The Yokogawa CSU-X1 spinning disk confocal scan head with very cool (-85 to -90°C) temperature, short exposure time Andor iXon-3 EMCCD camera on Nikon Eclipse Ti microscope system with PFS. *The Andor iXon-3 camera with EMCCD chip has $16 \times 16 \mu m$ pixels. At $100 \times$ magnification, it forms an image at the camera sensor with 160 nm pixel. While high sensitivity of EMCCD cameras has the advantage of potentially very low photobleaching.*
3. Computer set up with Andor IQ3 acquisition software for live cell imaging.

This software handles acquisition, processing, measurement, and analysis for various applications. The details of Andor IQ3 software available in the online available user guide. Since live amoebic cells are sensitive to exposure of light; it causes cells to move away from light and amoebae also acquire non-pseudopodia round cell. This gives an indication that the shortest possible exposure time and the minimum intensity of the fluorescence excitation light with neutral-density and UV filters and shutters are prime requirements for the optimal live cell imaging. The exposure time depends on the expression of the EYFP fusion protein and has to be determined empirically.

4. Mount 35mm glass base petri plate having DAPI and MitoRed stained EYFP fusion transformed amoebic cells on standard stage holder on the above-mentioned system.

Visualize Growing Cells

1. Transformed EYFP fusion protein containing *Dictyostelium discoideum* amoebic cells grown in HL5 media having 50-200µg/ml Geneticin-G418; up to log phase or on bacterial lawns.
2. Count cells by haemocytometer and dilute them up to the density of~1×10^6 cells/ml by HL5 media and pour in the glass base 35mm petri dishes.
3. Allow cells to attach for 10 min.
4. Add blue fluorescent nuclear stains 4', 6-diamidino-2-phenylindole (DAPI) diluted in deionized water (dH$_2$O) in concentration 1-2µg/ml and incubate for 10-30 min with wrapping in aluminium foil to protect exposure from light. Change the HL5 media after incubation 2-3 times to remove the stain.

Note:*1. DAPI binds in minor groove with AT cluster of dsDNA and produces ~20 times fluorescence enhancement by a xenon or mercury-arc lamp or with a UV laser; excitation maxima at 358 nm, and emission maxima at 461 nm (Tanious FA, Veal JM, Buczak H, Ratmeyer LS, Wilson WD., 1992 [26]).*

5. *If transformed cells are not staining nucleus efficiently with DAPI due to a higher level of overexpressed protein; then the addition of 1-2µl of DMSO into the diluted solution of DAPI before adding it with the cells enhances uptake of DAPI by cells.*

Caution:*DAPI is a known mutagen and should be handled with care. The dye must be disposed of safely and in accordance with applicable local regulations.*

6. After DAPI staining, washed cells with HL5 medium, were stained with 5-50nM of 9-[2-(4'-Methylcoumarin-7'-oxycarbonyl) phenyl]-3,6-bis (diethyl-amino) xanthylium chloride (MitoRed- Rhodamine based dye) Probes in the 35mm glass base petri plate and incubate for 5-30 min in the dark by wrapping

with aluminium foil. After staining, change HL5 medium 2-3 times to remove the extra stain. Observe cells using spinning disk confocal microscope (Poot M, Zhang YZ, Krämer JA, Wells KS, Jones LJ, Hanzel DK, Lugade AG, Singer VL,Haugland RP.,1996 [27]).

Note: *The fluorescent stains tetramethylrhodamine and rhodamine 123 accumulates in active mitochondria through passive diffusion. The longer incubation period exhaust cells which give the rounded appearance to the cells; without pseudopodia.*

Image acquisition of EYFP fusion protein

1. Mount 35mm glass base petri plate on the stage of Nikon Eclipse Ti inverted microscope and the Andor IQ3 acquisition software was set on time-lapse mode.
2. **Note:***The 63x or 100x oil immersion objectives gives better resolution of GFP fusion transformed Dictyostelium discoideum amoebae having 10 to 20μM diameter.*
3. Acquisition of images done at 1-2 frames per sec (FPS) by exciting EYFP with 488nm λ and detecting the emission at 545nm λ to localize the fusion protein; excitation of DAPI with 358nm λ and detection of emission of 461nm λ for the localization of nucleus. The excitation of MitoRed dye with 560nm λ and detection by the emission of 580nm λ for the localization of mitochondria.
4. Imaging was done with increasing concentration of MitoRed from 5nM to 50nM as well as with increasing incubation time from 5-20 min.
5. Acquisition of data completed depending on incubation time of stains in EYFP fusion protein expressing amoebae.
6. Repeat this process with a different set of DAPI and MitoRed stain concentration and incubations periods.

Evaluation of Data

The software Andore IQ3, user guide web address, http://www.imm.fm.ul.pt/wiki/lib/exe/fetch.php?media=bioimaging:iq_user_guidejune_2010_w.pdf [28] gives detailed aspects of the data acquisition and evaluation. This software offers like intensity measurements of a defined region of interests (ROIs). This also provides several plug-ins which may be useful in data evaluation. This tool is suitable for live cell image acquisition of amoeboid cells with various study aspects like lateral motility and study of fluorescence labelled microtubules.

DISCUSSION

In this live cell imaging protocol the focus was primarily to check; whether fluorescent constituent compounds of medium HL5 in the case of *Dictyostelium discoideum* can be used for osmolarity sensitive transformed constitutively expressing EYFP fusion protein strains. Our practical experience shows that; normally HL5 medium constituent gives a very high level of background fluorescence with fluorescent microscopes. While, this problem of HL5 medium fluorescence can be overcome up to the marked level by using spinning disk confocal microscopes with highly sensitive Yokogawa CSU-X1 spinning disk confocal scan head with very cool (-85 to -90°C) temperature, short exposure time Andor iXon-3 EMCCD camera.

In this protocol, both images of transformed live cell amoebic cells with DAPI and MitoRed stains (Figs. **3** and **4**) acquired within HL5 medium incubation and absence of background fluorescence caused by HL5 medium gives a clear indication of better sensitivity as well as the efficiency of spinning disk confocal microscopes over the normal fluorescence microscopes.

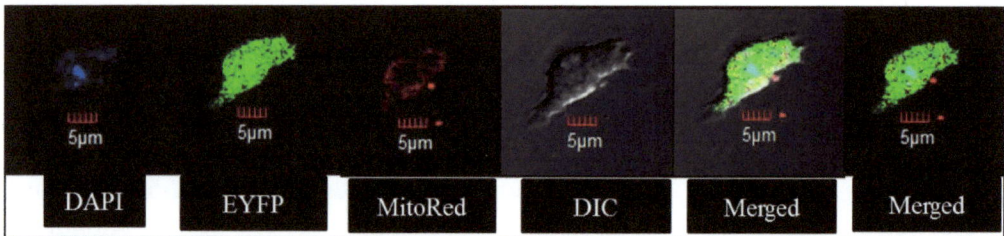

Fig. (3). Live cell imaging of EYFP fusion transformed amoebic cells, constitutively expressing EYFP fusion protein.The cells were stained with DAPI and MitoRed for nuclear and mitochondrial localization respectively. The DIC, DAPI, FITC and TRITC filters were used for images acquisition as mentioned below the image panels.

The sensitivity and precision of acquisition are so high that it can help in the avoidance of the use of an antifade agent. Moreover, this can help in the image acquisition in the live cells before photo-bleaching. However, by adjusting the optimal setting for image acquisition, through increasing Electron Multiply (EM) gain with decreasing exposure time with Andore IQ3 software, the photo-bleaching and other damages to the live cells during image acquisition can be reduced and images of motile amoebic cells with pseudopodia can be captured.

The 60x or 63x objective lens is a better choice for amoebic live cell imaging compared to 100x because higher power of objective lens causes more photo-bleach and reactive species damage.

In the live cell imaging of amoebic cells selection of optimal refractive index medium (glass, emersion oil) as well as optimal growth conditions (HL5 medium and 22°C temperature) are primary requirements to reduce the pressure and harm on live cells caused by bright white light and different high energy lasers; because of this HL5 medium was used in this protocol during image acquisition in place of PBS and KK2 buffers.

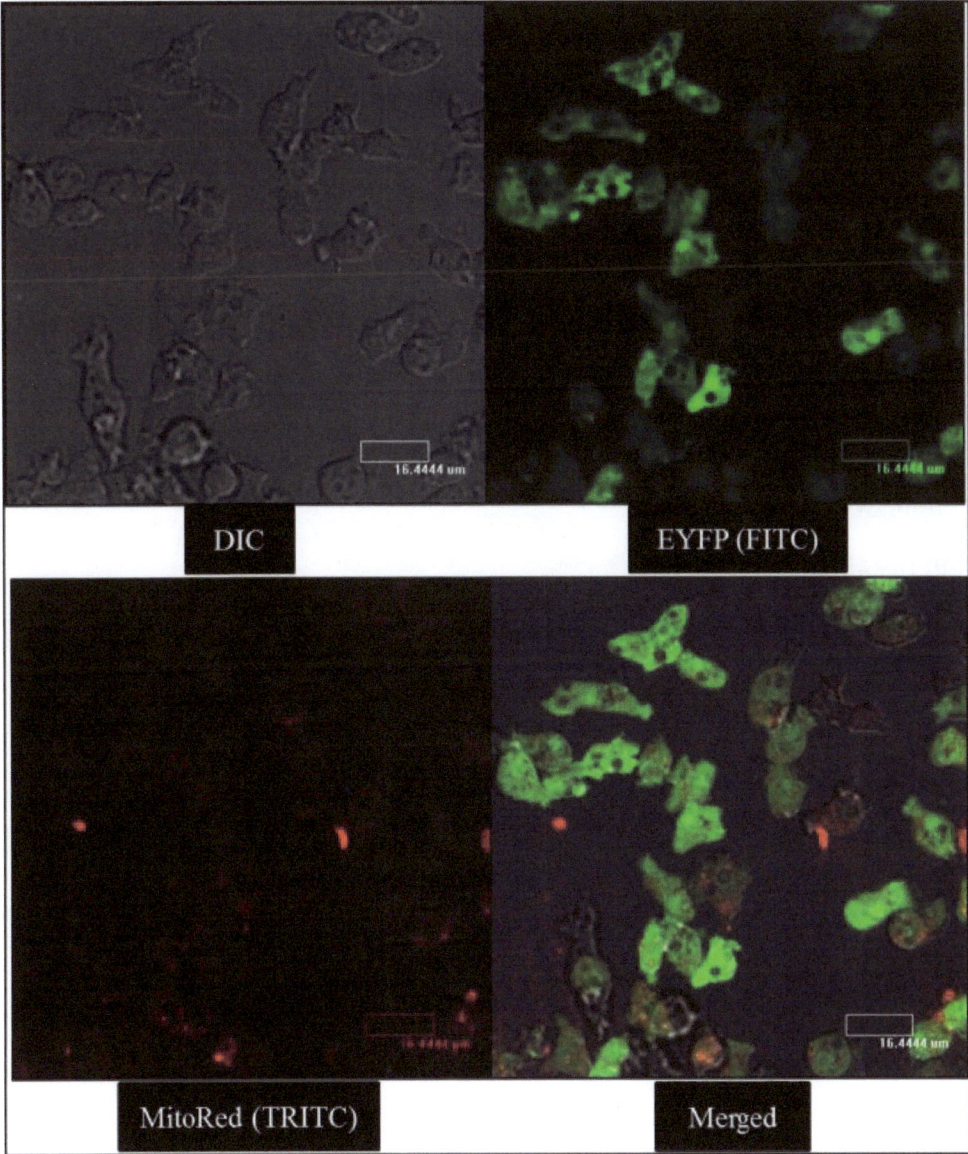

Fig. (4). Live cell images of transformed amoebic cells, constitutively expressing EYFP fusion reporter protein, stained with MitoRed for mitochondrial localization. The images were acquired with DIC, FITC and TRITC filters as mentioned below the image panel.

The live cell imaging movies for amoebic cell movements in response to light can be captured efficiently by selecting EM to gain at 50 and 1-3 frames per sec (FPS) with above Andore IQ3 software and EMCCD camera with Yokogawa spinning disk confocal head at the 60x objective lens.

So, in this way, the efficiency of live cell imaging in growth medium will give more authenticated and improved results in my opinion for the cutting edge science and will help in the improved level of live cell study of osmolarity sensitive strains.

Medium and Buffers

HL5 (for 1 litre)

Proteose peptone 14.3 gm

Yeast extract 7.15gm

Glucose 16gm

$Na_2HPO_4.2H_2O$ 0.626gm

KH_2PO_4 0.485gm

The pH was adjusted to 6.5, with dilute HCl and the volume made up with double distilled water.

SM agar

- 41.7 g SM agar

- Add ultrapure dH_2O to 1,000 ml and autoclaved to sterilise.

KK2 buffer

2.24 g KH_2PO_4

0.52 g K_2HPO_4, anhydrous

Add ultrapure dH_2O to 1,000 ml and autoclaved to sterilise in the 1,000 ml Duran bottles at 121 °C for 15 min. The pH after autoclaving should be ~6.1.

Add 2 ml of sterile 1 M $MgSO_4$ and stored at 22 °C.

Dense Suspension of *Klebsiella aerogenes* streaked out on SM agar plate and individual colonies renewed every 2 months from frozen stocks. A colony picked with a flame sterilised platinum loop or with a disposable sterile plastic needle or plastic loop and use to seed a universal bottle containing SM broth. Incubate for 48 h at 22°C and vortex before using and store it in the dark at 8°C as well as renew it every month.

Electroporation Buffer E50

2.38 g HEPES

1.86 g KCl

0.29 g NaCl

0.5 ml of 1 M $MgSO_4$

0.21 g $NaHCO_3$

0.08 g $NaH_2PO_4.2dH_2O$

Add ultrapure 500ml dH_2O and adjust pH 7.0 with KOH then sterilise with a 0.22μM filter using 500 ml Stericup filter unit and store at 4°C.

6. 20 mg/ml G418

552 mg Geneticin® G418 sulphate adjusted for the potency of 724 μg/mg by adding ultrapure dH_2O to 20 ml, Filter sterilised with 0.22 μM syringe filter and store at -20°C in aliquots.

Freezing Storage Medium

150 ml horse serum

11.25 ml [7.5% (v/v)] DMSO

Filter sterilise and aliquot into sterile 50 ml tubes and stored at -20°C and should be used within 1 year.

CONSENT FOR PUBLICATION

Not applicable.

CONFLICT OF INTEREST

The author confirms that this chapter contents have no conflict of interest.

ACKNOWLEDGEMENTS

Work in the laboratory supported by the CSIR-JRF and SRF fellowship and the Central Instrumentation Facility (CIF) of the School of Life Sciences, Jawaharlal Nehru University, New Delhi, India. Figure 2, 3 and 4 were produced by author in the lab during study.

REFERENCES

[1]   Eichinger L, Pachebat JA, Glöckner G, *et al.* The genome of the social amoeba Dictyostelium discoideum. Nature 2005; 435(7038): 43-57.
[http://dx.doi.org/10.1038/nature03481] [PMID: 15875012]

[2]   Brown   D,   Strassmann   JE.   CC3.0   copyright.
http://www.dictybase.org/Multimedia/DdLifeCycles/index.html

[3]   Loomis WF Jr. Temperature-sensitive mutants of *Dictyostelium discoideum.* J Bacteriol 1969; 99(1): 65-9.
[PMID: 5816728]

[4]   Williams JG. Transcriptional regulation of Dictyostelium pattern formation. EMBO Rep 2006; 7(7): 694-8.
[http://dx.doi.org/10.1038/sj.embor.7400714] [PMID: 16819464]

[5]   Li G, Alexander H, Schneider N, Alexander S. Molecular basis for resistance to the anticancer drug cisplatin in Dictyostelium. Microbiology 2000; 146(Pt 9): 2219-27.
[http://dx.doi.org/10.1099/00221287-146-9-2219] [PMID: 10974109]

[6]   Steinert M, Heuner K. Dictyostelium as host model for pathogenesis. Cell Microbiol 2005; 7(3): 307-14.
[http://dx.doi.org/10.1111/j.1462-5822.2005.00493.x] [PMID: 15679834]

[7]   Sussman M. Cultivation and synchronous morphogenesis of Dictyostelium under controlled experimental conditions. Methods Cell Biol 1987; 28: 9-29.
[http://dx.doi.org/10.1016/S0091-679X(08)61635-0] [PMID: 3298997]

[8]   Hirst J, Kay RR, Traynor D. *Dictyostelium* Cultivation, Transfection, Microscopy and Fractionation. Bio Protoc 2015; 5(11): 1485.
[http://dx.doi.org/10.21769/BioProtoc.1485] [PMID: 26167517]

[9]   Nellen W, Firtel RA. High-copy-number transformants and co-transformation in Dictyostelium. Gene 1985; 39(2-3): 155-63.
[http://dx.doi.org/10.1016/0378-1119(85)90309-9] [PMID: 4092928]

[10]   Manstein DJ, Schuster HP, Morandini P, Hunt DM. Cloning vectors for the production of proteins in *Dictyostelium discoideum.* Gene 1995; 162(1): 129-34.
[http://dx.doi.org/10.1016/0378-1119(95)00351-6] [PMID: 7557400]

[11]   Hughes JE, Kiyosawa H, Welker DL. Plasmid maintenance functions encoded on *Dictyostelium discoideum* nuclear plasmid Ddp1. Mol Cell Biol 1994; 14(9): 6117-24.
[http://dx.doi.org/10.1128/MCB.14.9.6117] [PMID: 8065344]

[12]   Parent CA, Devreotes PN. Molecular dissection of G protein-mediated signal transduction using random mutagenesis in *Dictyostelium.* Adolph KW, Ed Microb Genome Methods. CRC Press 1996; p. 1-13.

[13]   Parent CA, Devreotes PN. Molecular genetics of signal transduction in Dictyostelium. Annu Rev Biochem 1996; 65: 411-40. b
[http://dx.doi.org/10.1146/annurev.bi.65.070196.002211] [PMID: 8811185]

[14]   Chung CT, Niemela SL, Miller RH. One-step preparation of competent Escherichia coli: transformation and storage of bacterial cells in the same solution. Proc Natl Acad Sci USA 1989; 86(7): 2172-5.
[http://dx.doi.org/10.1073/pnas.86.7.2172] [PMID: 2648393]

[15]   Gaudet P, Pilcher KE, Fey P, Chisholm RL. Transformation of *Dictyostelium discoideum* with plasmid DNA. Nat Protoc 2007; 2(6): 1317-24.
[http://dx.doi.org/10.1038/nprot.2007.179] [PMID: 17545968]

[16]   Chalfie M, Tu Y, Euskirchen G, Ward WW, Prasher DC. Green fluorescent protein as a marker for gene expression. Science 1994; 263(5148): 802-5.
[http://dx.doi.org/10.1126/science.8303295] [PMID: 8303295]

[17]   Shaner NC, Patterson GH, Davidson MW. Advances in fluorescent protein technology. J Cell Sci 2007; 120(Pt 24): 4247-60.
[http://dx.doi.org/10.1242/jcs.005801] [PMID: 18057027]

[18]   Shimomura O, Johnson FH, Saiga Y. Extraction, purification and properties of aequorin, a bioluminescent protein from the luminous hydromedusan, Aequorea. J Cell Comp Physiol 1962; 59: 223-39.
[http://dx.doi.org/10.1002/jcp.1030590302] [PMID: 13911999]

[19]   Zimmer M. GFP: from jellyfish to the Nobel prize and beyond. Chem Soc Rev 2009; 38(10): 2823-32.
[http://dx.doi.org/10.1039/b904023d] [PMID: 19771329]

[20]   Conchello JA, Lichtman JW. Optical sectioning microscopy. Nat Methods 2005; 2(12): 920-31.
[http://dx.doi.org/10.1038/nmeth815] [PMID: 16299477]

[21]   Tanaami T, Otsuki S, Tomosada N, Kosugi Y, Shimizu M, Ishida H. High-speed 1-frame/ms scanning confocal microscope with a microlens and Nipkow disks. Appl Opt 2002; 41(22): 4704-8.
[http://dx.doi.org/10.1364/AO.41.004704] [PMID: 12153106]

[22]   Wang E, Babbey CM, Dunn KW. Performance comparison between the high-speed Yokogawa spinning disc confocal system and single-point scanning confocal systems. J Microsc 2005; 218(Pt 2): 148-59.
[http://dx.doi.org/10.1111/j.1365-2818.2005.01473.x] [PMID: 15857376]

[23]   Conchello JA, Lichtman JW. Theoretical analysis of a rotating-disk partially confocal scanning microscope. Appl Opt 1994; 33(4): 585-96.
[http://dx.doi.org/10.1364/AO.33.000585] [PMID: 20862053]

[24]   Frigault MM, Lacoste J, Swift JL, Brown CM. Live-cell microscopy - tips and tools. J Cell Sci 2009; 122(Pt 6): 753-67.
[http://dx.doi.org/10.1242/jcs.033837] [PMID: 19261845]

[25]   Müller-Taubenberger A, Ishikawa-Ankerhold HC. Fluorescent reporters and methods to analyze fluorescent signals.Eichinger L, Rivero F, Eds Dictyostelium discoideum Protocols, Sec Edition, Methods Mol Biol. Humana Press 2013; 983: p. 93-112.
[http://dx.doi.org/10.1007/978-1-62703-302-2_5]

[26]   Tanious FA, Veal JM, Buczak H, Ratmeyer LS, Wilson WD. DAPI (4′,6-diamidino-2-phenylindole) binds differently to DNA and RNA: minor-groove binding at AT sites and intercalation at AU sites. Biochemistry 1992; 31(12): 3103-12.
[http://dx.doi.org/10.1021/bi00127a010] [PMID: 1372825]

[27]   Poot M, Zhang YZ, Krämer JA, *et al.* Analysis of mitochondrial morphology and function with novel fixable fluorescent stains. J Histochem Cytochem 1996; 44(12): 1363-72.
[http://dx.doi.org/10.1177/44.12.8985128] [PMID: 8985128]

[28]   http://www.imm.fm.ul.pt/wiki/lib/exe/fetch.php?media=bioimaging:iq_user_guide_june_2010_w.pdf

CHAPTER 14

The Recent Advancement in Rapid Golgi Method and Result Interpretation

Surya Prakash Pandey[1,*], Mallikarjuna Rao Gedda[2] and Abhishek Pathak[1]

[1] Department of Neurology, Institute of Medical Sciences, Banaras Hindu University, Varanasi, India

[2] Department of Biochemistry, Institute of Science, Banaras Hindu University, Varanasi, India

Abstract: The foundation knowledge of recent advancements of neuroscience was based on the Golgi staining observations. This is one of the best approaches to visualise the neuronal cytoarchitecture and complete morphology of neurons with incomparable clarity. This technique is based on the principle of heavy metal impregnation. There are many modifications and advancement occurred to improve the visualization. This chapter will provide the recently used protocols to visuals the neuronal architecture, dendritic arborization and spine density in different brain regions. Along with the manual observation, the present chapter also describes the currently used tools and software for the better understanding and visualisation of neurons.

Keywords: Brain, Dendritic branching, FIJI tool, Golgi staining, Metal, Neuronal morphology, Neuron, Spine count, Sholl analysis, Software, Transcardial perfusion.

INTRODUCTION

The Golgi method was first introduced by Camillio Golgi [1] but initially, its procedure was very lengthy. Therefore, many modifications and advancement occurred to improve visualization. These variations were focused for better staining procedure by reducing the time for staining, decreasing precipitation, endorsing identical crystallization, increasing consistency and reproducibility of neurons by the rapid Golgi method.

USE

This chapter will provide the recently used protocols to visuals the neuronal architecture, dendritic arborization and spine density in different brain regions.

* **Corresponding author Surya Prakash Pandey:** Department of Neurology, Institute of Medical Sciences, Banaras Hindu University, Varanasi, India; E-mail: suryabhu2009@gmail.com

Sandeep Kumar & Dhiraj Kumar (Eds.)

Along with the manual observation, the present chapter also describes the currently used tools and software for the better understanding and visualization of neurons.

- Neuronal architecture in a different region of the brain (deeper and cortical part of the brain)
- Dendritic arborization (Apical and basal) and length
- Spine count or spine density
- Cell body morphology

GENERAL PRINCIPLES OF THE GOLGI METHODS

The principle behind the Golgi technique is to heavy metal impregnation which forms a large complex of precipitate (lipoprotein-chromic silver complex). The precipitate further not crosses the membrane. Furthermore, the densely impregnated cells with cell architecture can be observed under a light microscope. This stain has some specific property [2].

- Very few neuronal cells (1 to 5%) have metallic deposition out of a large number of brain cells and the unstained part gives clear and transparent background.
- Rapid Golgi does not show gradations in staining; the CNS cells are either completely opaque to light or perfectly transparent.

COMPOSITION OF RAPID-GOLGI FIXATIVE

Fixative prepared by mixing of Potassium dichromate (5 gram), Chloral hydrate (5 gram), Glutaraldehyde (8 ml), Formaldehyde (6 ml) and Dimethyl sulfoxide (200 μl) in 100 ml of distal water.

Impregnation solution (0.75% Silver nitrate):

750 mg of silver nitrate ($AgNO_3$) dissolved in 100 ml distilled water. (Avoid light; due to light sensitive)

Normal saline

0.9% of sodium chloride (NaCl)

RAPID GOLGI STAINING FOR SMALLER BRAIN TISSUE OR CORTICAL BRAIN

Tissue Processing: the whole brain of Animal should be isolated in deep anaesthetized condition by decapitated quickly. For mouse brain or smaller brain no need for further dissection whereas, rat brain requires four pieces (about 1 cm^3) for proper impregnation. Then tissues are kept in an amber coloured bottle with a sufficient amount of fixative. The description of tissue processing as described below:

- 1st day: The old fixative is poured and freshly prepared Golgi fixative added. The bottle should be closed tightly with a lid and stored in the refrigerator.
- 2nd day: The fixative in which the tissue was kept in the previous day is decanted. The tissues rinsed by adding a small amount of fixative thereafter, remaining fixative (old previous day prepared fixative) is poured into the bottle and kept again in the dark chamber.
- 3rd day: Fresh solution of fixative prepared. The old fixative is poured out and brain tissue is rinsed once or twice and fresh fixative is poured into the tissue containing bottle and again kept in the refrigerator.
- 4th day: The brain tissue remained undisturbed.
- 5th day: The tissue blocks are rinsed several times in 0.75% aqueous solution of silver nitrate (AgNO$_3$) till the reddish brown colour of the potassium dichromate-silver complex disappeared. Tissues are then placed in a 0.75% AgNO$_3$ solution and kept in the dark for a minimum of 48 hours.
- Tissue embedding: After 48 hours, the tissue pieces are placed in a petri dish and the silver deposits were brushed off gently. The tissue is blotted and then dehydrated in absolute alcohol for 10 min. After dehydration, the tissue blocks are carefully mounted on the required plane of cutting on to block holders. After that, the block holder is placed onto a vibratome sliding/sledge microtome.
- Sectioning: 100 to 120 mm thick sections are taken in submerged chilled 70% alcohol by vibratome or sledge or sliding microtome. The sections are collected in a petri dish containing 70% alcohol.
- Clearing and mounting: The sections are first dry on blotting paper then moved in an alcohol-xylene mixture (1:1). Finally placed in xylene containing petri dish for clearing. The completion of the clearing step was confirmed, once the sections sank completely and also when the sections transformed to translucent. The sections are then mounted on slides with DPX and coverslipped. The slides are air dried at room temperature for a week.

MODIFIED RAPID GOLGI STAINING SUBCORTICAL REGIONS (HIPPOCAMPUS)

The fixative requires more time for proper impregnation in the deep brain or sub-cortical brain regions and time further increases in dead tissue. Because this modification approach delivers the fixative to the deep brain by blood capillaries in alive stage gives a better result and less time of impregnation [3].

Mice are transcardially perfused (described in last) by Golgi Fixative followed by normal saline and brain was dissected out quickly. Thereafter, dissected brains were post-fixed in Golgi fixative solution for 2 days at 4°C. After fixation, tissue was washed and impregnated with 0.75% silver nitrate for 48 hours in the dark. 100 um thick sections were cut by vibratome and collected in chilled 70% alcohol. The sections were dehydrated in absolute alcohol, cleared in xylene and mounted in DPX. Slides were dried at 37°C for 3-4 days.

MICROSCOPY Z STACKING

Microscopy can be done by any light microscope but neuronal arbours are extended in the depth of sections so only one plan can't give the complete picture. This problem can be overcome by using either of the two following attachment in a microscope. 1) With the help Camera Lucida, we can draw the image of a complete neuron on the paper by tracing all branches. Or 2) by using the Z-stacking stage in the microscope we can take many pictures of different plans and by merging of these gives us a three-dimensional complete picture of the neuron. The purpose of these two approaches is to analyse complete arborisation in one plane. For dendritic arbour study 40X and for spine study 100X picture should be taken.

SHOLL ANALYSIS

The Sholl analysis application is freely available in Fiji tool (http://fiji.sc/). Sholl is a quantitative analysis method commonly used to illustrate the morphological characteristics of an imaged neuron (Fig. **1**). In this method,the neuron is initially radially divided lengthwise from the centre of the cell body and calculates the branch point and branching of dendrites. This will give Dendritic arborisation (Apical and basal), dendritic length and total branch point [4].

MANUAL COUNTING OF SPINE AND DENDRITE

Martin L. Feldman and Alan Peters (1979) [5] describe in detail to evaluate the spine density and spine count manually. They used geometrical criteria because from one side of the section we can see whole spine arrangements on the diameter of the dendrite, therefore, this gives a certain accuracy to the total number of the spine.

$$N = \frac{n\,m\,[(Dr + S1)^2 - (Dr + Sd)^2]}{[\Theta\pi/90(Dr + Sl)^2] - 2\,[(Dr + S1)\sin\Theta\,.\,(Dr + Sd)]}$$

Where

Dr, the radius of dendrite

S1, length of the spine

Sd, the diameter of the spine head

n, number of spines visible

N, the estimated value of N

As, area of dendritic spine field

in which spine tips are counted

Θ, the angle from the axis

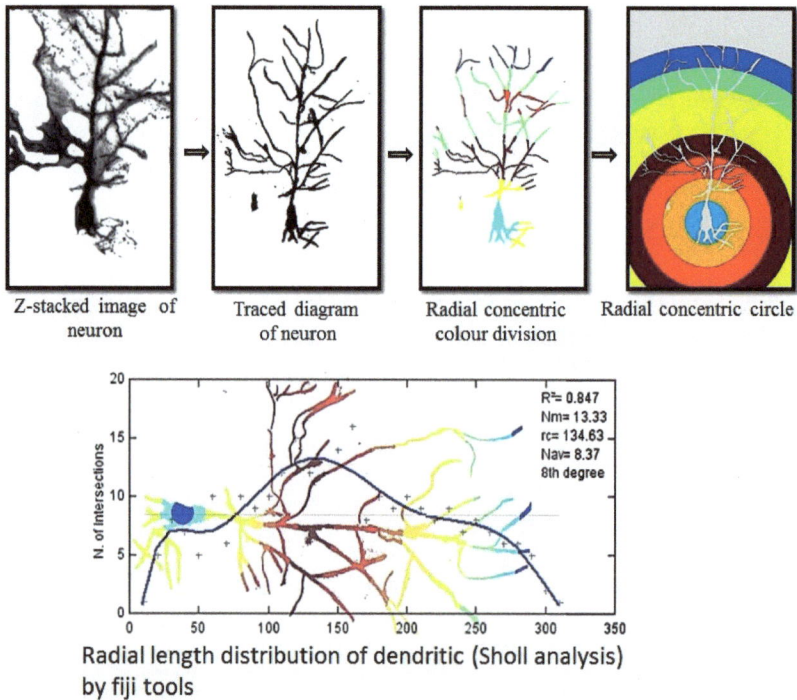

Z-stacked image of neuron   Traced diagram of neuron   Radial concentric colour division   Radial concentric circle

Radial length distribution of dendritic (Sholl analysis) by fiji tools

Fig. (1). This diagram represents the procedure and outcome of Sholl analysis.

PRECAUTION

- The impregnation time for a particular animal and tissue should be standardized.
- Silver nitrate is light sensitive so prepare the fresh solution and kept in dark or covet it with aluminium foil.

- Before adding silver nitrate proper washing with a brush is required in silver nitrate solution many times.
- Collect tissue sections in chilled 70% to avoid fragmentation of section.

TRANSCARDIAL PERFUSION

Transcardial perfusion under terminal anaesthesia is a method commonly used for tissue fixation in histochemical localization protocols. This method takes advantage of the subject's circulatory system to deliver the fixative solution evenly throughout the body tissues with optimal penetration of the brain. Fixation preserves the ultrastructure and stabilizes the protein and peptide conformation so that antibodies can bind to antigen sites.

The perfusion unit contained two cylinders- like containers hanged at the height of 120 cm to create pressure equal to rat heart. Container 'A' was filled with normal saline and container 'B' was filled with Golgi solution. Mice were anaesthetized with anaesthetic ether vapours in desiccators. The mice were placed on the rack and an incision was made with a scalpel through the abdomen to the length of the diaphragm. With sharp scissors, the connective tissue was cut at the bottom of the diaphragm to allow access to the rib cage. With large scissors, (blunt side down) the ribs were cut through the midline of the rib cage to open up thoracic cavity. While holding heart steady, insert the catheter needle directly into protrusion of left ventricle to extend straight up. The needle position was secured by clamping in place. The cut was made in the right atrium with sharp scissors, so as to replace the whole blood with normal saline. When the whole blood was cleared from the body (≈30mls for the small animal), the Golgi solution (≈50mls) was allowed to run through the body. Care was taken not to introduce air. Spontaneous movement (formalin dance) and lightened colour of the liver are good indicators of the clearing. The animal was removed from the rack, and decapitated as close as possible to the back of the ears. An incision was made with the scalpel from the base of the head up to the eyes. The skin was pulled away from the skull, and the skull plates were removed to expose the brain. With the flat end of the spatula pushed under the brain, the optic, olfactory, and cranial nerves were removed. The brain was removed out and placed in Golgi fixative filled specimen bottle and postfixed at 4°C [6].

CONSENT FOR PUBLICATION

Not applicable.

CONFLICT OF INTEREST

The author confirms that this chapter contents have no conflict of interest.

ACKNOWLEDGEMENTS

Declared none.

REFERENCES

[1] Golgi C. Sulla struttura della sostanza grigia del cervello. *Gazzetta medica italiana.* Lombardia 1873; 33: 244-6.

[2] Jacobs B, Driscoll L, Schall M. Life-span dendritic and spine changes in areas 10 and 18 of human cortex: a quantitative Golgi study. J Comp Neurol 1997; 386(4): 661-80.
[http://dx.doi.org/10.1002/(SICI)1096-9861(19971006)386:4<661::AID-CNE11>3.0.CO;2-N] [PMID: 9378859]

[3] Pandey SP, Prasad S. Diabetes mellitus type 2 induces brain ageing and memory impairment in mice: neuroprotective effects of Bacopa monnieri extract Topics in biomedical gerontology. Springer 2017.

[4] Sharma HR, Thakur MK. Correlation of ERα/ERβ expression with dendritic and behavioural changes in CUMS mice. Physiology & behavior 2015; 1;145: 71-83 .

[5] Shankaranarayana Rao BS, Raju TR, Meti BL. Long-lasting structural changes in CA3 hippocampal and layer V motor cortical pyramidal neurons associated with self-stimulation rewarding experience: a quantitative Golgi study. Brain Res Bull 1998; 47(1): 95-101.
[http://dx.doi.org/10.1016/S0361-9230(98)00056-2] [PMID: 9766395]

[6] Sharma HR, Thakur MK. Correlation of ERα/ERβ expression with dendritic and behavioural changes in CUMS mice. Physiol Behav 2015; 145: 71-83.
[http://dx.doi.org/10.1016/j.physbeh.2015.03.041] [PMID: 25837835]

CHAPTER 15

Molecular Markers for the Evaluation of Clonal Fidelity in Medicinal Plants

Arpan Modi and **Surapathrudu Kanakala**[*]

Institute of Plant Sciences, Agricultural Research Organization, Volcani Center, Israel

Abstract: Medicinal plants are major sources of secondary metabolites for which they have been paid more attention by pharmaceutical industries. In order to produce these secondary metabolites, medicinal plants are cultivated and for that plant tissue or organ, culture can be a suitable alternative. However, these plants are treated with plant hormones and elicitors to enhance the secondary metabolites and such elicitation may lead to genetic or epigenetic changes which are known as somaclonal variations. Thus, a stringent method of monitoring is required to observe the true-to-types of these medicinal plants when multiplied through tissue culture. Molecular markers like Randomly Amplified Polymorphic DNA (RAPD), Inter-Simple Sequence Repeat (ISSR), and Simple Sequence Repeats (SSR) are highly suitable markers to assess clonal fidelity in micropropagated medicinal plants. In the present chapter, the execution of such markers to check somaclonal variations in tissue culture raised medicinal plants is discussed in detail.

Keywords: Clonal fidelity, ISSR, Medicinal plants, Micropropagation, Molecular markers, Organ culture, Plant tissue, RAPD, SSR, Simple sequence repeats, Somaclonal variations, Tissue culture, True-to-type plants.

INTRODUCTION

Plant tissue culture technique can be a good source of large scale plant production for dioecious plants [1 - 3], endangered plants [4 - 6], medicinal plants [7] and plants showing poor germination [8, 9]. The techniques are completely suitable for the propagation of medicinal plants, demands of which are increasing everyday by pharmaceutical industries, due to their properties to synthesize highly valuable secondary metabolites. Plant tissue culture or micropropagation technique has several advantages over conventional ways of propagation which include shorter time and space to produce population from very less starting material, the feasibility of plant production throughout the year, production of

[*] **Corresponding author Surapathrudu Kanakala:** Institute of Plant Sciences, Agricultural Research Organization, Volcani Center, Israel; E-mail: kanakalavit@gmail.com

Sandeep Kumar & Dhiraj Kumar (Eds.)

disease-free population and high multiplication rate. Various medicinal plants, propagated through tissue culture for commercial production of secondary metabolites are enlisted in Table **1** [10].

Somaclonal variations are likely to occur in plant tissue culture, especially attended through callus culture. Somaclonal variations may come through selection of explant (highly differentiated tissue such as leaf, stem and root are more prone to give somaclonal variation than pre-existing meristems like shoot tip and axillary bud), mode of regeneration (callus cultures), culture period and subculture cycles, culture environment such as growth regulators, temperature, light, osmolarity and agitation rate of liquid culture. Of course, genotype and ploidy level of the plant have a greater influence on somaclonal variations to occur [23].

Table 1. Micropropagated medicinal plants their secondary chemicals and uses.

Sr. No.	Plant	Secondary Metabolite	Uses	Reference
1	*Catharanthus roseus*	Ajmalicine, Vincristine, Vinblastine	Antihypertensive, Antileukemic	[11]
2	*Artemisia annua*	Artemisinin	Antimalarial	[12]
3	*Coptis japonica*	Berberine	Intestinal ailment	[13]
4	*Camptotheca acuminate*	Camptothecin	Antitumor	[14]
5	*Papaver somniferum*	Codeine, Morphine	Sedative	[15]
6	*Colchium autumnale*	Colchicine	Antitumor	[16]
7	*Digitalis lanata*	Digoxin	Heart stimulant	[17]
8	*Dioscorea deltoidea*	Diosgenin	Steroidal precursor	[18]
9	*Orchrosia elliptica*	Ellipticine	Antitumour	[19]
10	*Podophyllum petalum*	Podophyllotoxin	Antitumour	[20]
11	*Lithospermum erythrorhizon*	Shikonin	Antibacterial	[21]
12	*Taxus brevifolia*	Taxol	Anticancer	[22]

After the discovery of polymerase chain reactions, various molecular marker systems had been developed. These markers are used mostly in genotyping, sex determination, molecular breeding, varietal identification and clonal fidelity testing. PCR based molecular markers are simple, convenient, cost-effective and require a less laborious process. Compared with them, RFLP and AFLP (although PCR based) are more complicated and many times require the use of radioactive substances, expensive enzymes and extensive care [24].

MICROPROPAGATION PROTOCOL

The pre-requisite step for the clonal fidelity testing is the plant's propagation technique which includes many parameters. As mentioned by the study [23], explant selection is the first step for the initiation of good quality culture. Highly differentiated tissues like leaf, stems or roots are very prone to produce somaclonal variations. Use of meristematic tissue would be the best preference for this purpose. Even in the same tissue type, the different cell may have different ploidy level, the use of which may introduce the variations. Hypocotyl tissue is an example of having cells with different ploidy levels. However, there were reports involving such a mechanism when micropropagation was attempted through protoplast culture. Likewise, cultures resulted from an immature embryo would also produce somaclonal variants [25]. Mode of organogenesis plays a vital role in generating somaclonal variations. Organogenesis achieved through indirect route or from protoplast cultures may lead to induce genetic variations. The magnitude of stress on the explant, especially callus multiplying in an unorganized manner is higher as compared to meristem. As a whole, the tissue culture process activates the retrotransposons present in plant cell which possibly induce mutation in unorganized mitotic divisions. Hormones may induce abnormal mitoses in the micropropagation, however, many times, the timing of cell cycle is disturbed which may lead to abnormalities [26].

MOLECULAR MARKERS

For the evaluation of clonal fidelity in medicinal plants, several markers have been developed. Mainly there are two types of molecular markers *viz.,* biochemical and nucleic acid-based markers. Biochemical markers are mostly protein-based and secondary metabolites based (in case of medicinal plants) but they also show variations in response to environmental changes thus may not be helpful in the evaluation of true-to-typeness of the plant. In nucleic acid-based markers, DNA based markers are highly suitable for this purpose. PCR based markers like Randomly Amplified Polymorphic DNA (RAPD), Inter-Simple Sequence Repeats (ISSR), Simple Sequence Repeats, Sequence Cleaved Amplified Region (SCAR) are used often in the evaluation of somaclonal variations. In the present chapter, how they are employed (mainly RAPD, ISSR and SSR) to assess the clonal fidelity of medicinal plants are discussed in details.

The pre-requisite step towards the evaluation of clonal fidelity of micropropagated plants is the extraction of genomic DNA from the mother plant and tissue culture raised plants afterwards. The most efficient protocol for the extraction of genomic DNA was established by [27] and has been used widely for the DNA extraction from plant tissues. However, medicinal plants are rich in secondary metabolites, polysaccharides and polyphenols which co-precipitated with the genomic DNA and hinder the amplification in downstream processes. Polyphenols are released from the vacuole and counteracted by cellular oxidase. Later on, they react with nucleic acid and causing their oxidation to make them unsuitable for further reactions. Gelling polysaccharides prevent solubilization of DNA and increase the viscosity of the solution which thereby affects the electrophoretic mobility of the DNA during gel electrophoresis [28]. Thus, a modified version of universally accepted cetyl-trimethyl ammonium bromide (CTAB) protocol was established by several researchers [29 - 31]. These protocols mainly deal with all the hindrances present in the tissues of the medicinal plant and eliminate them to confirm the success of downstream processes.

Among PCR based molecular markers, RAPDs are widely used markers for several purposes. These are 10 nucleotides, dominant markers which bind randomly within sample DNA and with the use of polymerase chain reaction (PCR) the targeted regions got amplified [32]. However, due to its affinity to bind at non-specific sites, they sometimes produce false positive results and hence, may not give reproducible results. Still, they are widely used in all the crops including medicinal plants to evaluate somaclonal variations. On the other hand, Inter-Simple Sequence Repeats (ISSR) markers are semi-arbitrary, multi-loci single primer based markers targeting microsatellite regions. Unlike RAPD, they can be considered as reproducible markers [33]. Both the primers required screening, validation and confirmation of clonal fidelity steps. In the screening step, primers are selected from available sources of RAPD and ISSR with the mother plant. Primers showing more than 7, clear, distinct and reproducible amplicons are selected and then genomic DNAs obtained from 10% of the tissue culture raised population are subjected to amplify through screened primers along with mother plant DNA. Commonly used PCR condition for both RAPD and ISSR is mentioned in Table **2**. PCR reaction for both RAPD and ISSR may be carried out with 25 µl reaction volume containing 100 ng of genomic DNA, 3-5 units of DNA polymerase enzyme with respective buffer, 0.5 mM $MgCl_2$ (final concentration), 0.2 mM of dNTPs (final concentration), 2 µl of primer (10 pmol). Scoring of bands is carried out after running PCR products on agarose gel electrophoresis. The expected result of clonal fidelity testing is similar banding pattern of tissue culture raised clones with the banding pattern of the mother plant.

Using these markers, researchers were able to show no variations in numbers of plants species like *Stevia rebaudiana* [8, 34], *Guadua angustifolia* [35], *Canna indica* [36] and *Podophyllum hexandrum* [37]. In the assessment of genetic fidelity in *Swertia chirayita* conducted by the study [38], it was observed that tissue culture raised population showed 97% similarities when analyzed with

RAPD markers using Unweighted Pair Group Method with Arithmetic Mean (UPGMA). To check the potentiality of RAPD primers, the clonal fidelity in *Silybum marianum* (L.) with four different samples *viz.,* wild grown plants, seed derived plantlets, callus tissue and regenerated plantlets is studied [39]. They observed significant variations among these samples as amplified by RAPD primer OPC10, however, other 2 primers (OPC8 and OPC9) showed no polymorphism between selected samples.

Table 2. PCR conditions for DNA samples amplified through RAPD and ISSR markers.

Technique	Stage	Temperature (°C)	Time (mm:ss)	Repeat
RAPD	Initial denaturation	94	05:00	1
	Denaturation	94	01:00	
	Annealing	37	01:00	35-40
	Extension	72	02:00	
	Final extension	72	05:00	1
	Hold	4	∞	-
ISSR	Initial denaturation	94	05:00	1
	Denaturation	94	01:00	
	Annealing	40-55	01:00	35-40
	Extension	72	02:00	
	Final extension	72	05:00	1
	Hold	4	∞	-
SSR	Initial denaturation	94	05:00	1
	Denaturation	94	01:00	
	Annealing	$Tm - 2$	00:30	35-40
	Extension	72	01:00	
	Final extension	72	05:00	1
	Hold	4	∞	-

On the other hand, Simple Sequence Repeats (SSR) markers are highly specific and mainly target alleles. They amplify repeat sequences, frequently observed in the flanking region of the gene and are tightly linked. They are co-dominant markers having forward and reverse primers separately. However, they are very less used in the detection of clonal fidelity and mostly used in genotyping as well as marker-assisted breeding [24]. Cycling conditions for PCR amplification using SSR markers is mentioned in Table **2**. Whereas, PCR components used in this technique remain the same except 1 µl of each forward and reverse primers are taken.

To the date, no microsatellite markers have been employed to study the clonal fidelity of micropropagated medicinal plants. However, 1 bp difference in micropropagated guava plants, bearing medicinal properties, as revealed by amplification of DNA of both mother plant and clones through locus mPgCIR07 while amplification of both DNA with other 16 loci showed no genetic variation [40]. They proposed very fewer chances for the phenotypic difference as compared to the mother plant.

CONCLUSION

Micropropagation of medicinal plants warrants many benefits as compared to conventional ways of propagation. It is a pre-requisite step for metabolic engineering. With the technique itself, many secondary metabolites can be produced *in vitro* which saves space and time. Astringent method of detection or monitoring is also necessary to check the true-to-types of the plants. Medicinal plants may be subjected to genetic or epigenetic changes throughout the cycles of micropropagation. These changes may have an influence on the phenotypic characteristic of the plant and the outcome of the micropropagation protocol can be something unexpected. Monitoring can be done in several ways *viz.,* morphological and molecular markers. Morphological markers require time for the plant to mature while molecular markers can also be employed at the hardening stage also. Some biochemical markers are also available but they may also show false positive or negative results as they are under the influence of environmental changes. DNA markers are quite useful in this manner. RAPD and ISSR, being the simplest DNA markers and are widely used to study clonal fidelity in medicinal as well as horticultural crops. However, reproducibility of the result obtained is lesser in RAPD as compared to ISSR. Another DNA marker SSR is highly efficient and shows reproducible results. They are locus-specific and can be used to study clonal fidelity of micropropagated medicinal plants. Till date, these molecular markers have been employed to study the genetic variations created during tissue culture of medicinal plants. There is a need to develop and execute the molecular markers to study epigenetic changes made by micropropagation.

CONSENT FOR PUBLICATION

Not applicable.

CONFLICT OF INTEREST

The author confirms that this chapter contents have no conflict of interest.

ACKNOWLEDGEMENTS

Declared none.

REFERENCES

[1]   Reuveni O, Shelsinger DR, Lavi U. *In vitro* clonal propagation of dioecious *Carica papaya*. Plant Cell Tissue Organ Cult 1990; 20: 41-6.
[http://dx.doi.org/10.1007/BF00034755]

[2]   Bekheet S. Direct organogenesis of date palm (*Phoenix dactylifera* L.) for propagation of true-to-type plants. Sci Agric 2013; 4: 85-92.

[3]   Kumar N, Singh AS, Modi AR, *et al*. Genetic stability studies in micropropagated Date palm (*Phoenix dactylifera* L.) plants using microsatellite marker. J For Sci 2010; 26: 31-6.

[4]   Pandey P, Mehta R, Upadhyay R. *In vitro* propagation of an endangered medicinal plant *Psoralea corylifolia* Linn. Asian J of pharm and clin res 2013; 115-118.

[5]   Wala BB, Jasrai YT. Micropropagation of an endangered medicinal plant: *Cuculigo orchioides*. Plant Tissue Cult 2003; 13: 13-9.

[6]   Dang JC, Kumaria S, Kumar S, Tandon P. Micropropagation of *Ilex khasiana*, a critically endangered and endemic holly of Northeast India. AoB Plants 2011; 2011(May)plr012
[http://dx.doi.org/10.1093/aobpla/plr012] [PMID: 22476482]

[7]   Chaturvedi HC, Jain M, Kidwai NR. Cloning of medicinal plants through tissue culture--a review. Indian J Exp Biol 2007; 45(11): 937-48.
[PMID: 18072537]

[8]   Modi AR, Patil G, Kumar N, Singh AS, Subhash N. A simple and efficient *in vitro* mass multiplication procedure for *Stevia rebaudiana* Bertoni and analysis of genetic fidelity of *in vitro* raised plants through RAPD. Sugar Tech 2012; 14: 391-7.
[http://dx.doi.org/10.1007/s12355-012-0169-6]

[9]   Samuel K, Debashish B, Madhumita B, *et al*. *In vitro* germination and micropropagation of *Givotia rottleriformis* Griff. In Vitro Cell Dev Biol Plant 2009; 45: 466-73.
[http://dx.doi.org/10.1007/s11627-008-9181-7]

[10]   Rao SR, Ravishankar GA. Plant cell cultures: Chemical factories of secondary metabolites. Biotechnol Adv 2002; 20(2): 101-53.
[http://dx.doi.org/10.1016/S0734-9750(02)00007-1] [PMID: 14538059]

[11]   Hoopen HJG, Gulik WM, Schltmann AE, *et al*. Ajmalicine production by cell cultures of *Catharanthus roseus*: from shake flask to bioreactor. Plant Cell Tissue Organ Cult 1994; 38: 85-91.
[http://dx.doi.org/10.1007/BF00033865]

[12]   Nair MSR, Acton N, Klayman DL, Kendrick K, Basile DV, Mante S. Production of artemisinin in tissue cultures of Artemisia annua. J Nat Prod 1986; 49(3): 504-7.
[http://dx.doi.org/10.1021/np50045a021] [PMID: 3760887]

[13]   Morimoto T, Hara Y, Kato Y, *et al*. Berberine production by cultured *Coptis japonica* cells in a One-stage culture using medium with a high copper concentration. Agric Biol Chem 1988; 52: 1835-6.
[http://dx.doi.org/10.1271/bbb1961.52.1835]

[14]   Li Z, Liu Z. Camptothecin production in *Camptothecia acuminata* cultured hydroponically and with nitrogen enrichments. Can J Plant Sci 2005; 85: 447-52.

[http://dx.doi.org/10.4141/P04-020]

[15] Siah CL, Doran PM. Enhanced codeine and morphine production in suspended Papaver somniferum cultures after removal of exogenous hormones. Plant Cell Rep 1991; 10(6-7): 349-53.
[http://dx.doi.org/10.1007/BF00193157] [PMID: 24221672]

[16] Ghosh B, Mukharjee S, Jha TB, Jha S. Enhanced colchicine production in root cultures of *Gloriosa superba* by direct and indirect precursors of the biosynthetic pathway. Biotechnol Lett 2002; 24: 231-4.
[http://dx.doi.org/10.1023/A:1014129225583]

[17] Hagimori M, Matsumoto T, Obi Y. Studies on the production of *Digitalis* cardenolides by plant tissue culture. Plant Physiol 1982; 69(3): 653-6.
[http://dx.doi.org/10.1104/pp.69.3.653] [PMID: 16662267]

[18] Rokem JS, Tal B, Goldberg I. Methods for increasing diosgenin production by *Dioscoria* cells in suspension cultures. J Nat Prod 1985; 48: 210-22.
[http://dx.doi.org/10.1021/np50038a004]

[19] Kouadio K, Chenieux JC, Rideau M, Viel C. Antitumor alkaloids in callus cultures of *Ochrosia elliptica.* J Nat Prod 1984; 872-874.

[20] Giri A, Lakshmi Narasu M. Production of podophyllotoxin from *Podophyllum hexandrum*: a potential natural product for clinically useful anticancer drugs. Cytotechnology 2000; 34(1-2): 17-26.
[http://dx.doi.org/10.1023/A:1008138230896] [PMID: 19003377]

[21] Kim DJ, Chang HN. Enhanced shikonin production from *Lithospermum erythrorhizon* by in situ extraction and calcium alginate immobilization. Biotechnol Bioeng 1990; 36(5): 460-6.
[http://dx.doi.org/10.1002/bit.260360505] [PMID: 18595102]

[22] Ellis DD, Zeldin EL, Brodhagen M, Russin WA, McCown BH. Taxol production in nodule cultures of *Taxus.* J Nat Prod 1996; 59(3): 246-50.
[http://dx.doi.org/10.1021/np960104g] [PMID: 8882426]

[23] Krishna H, Alizadeh M, Singh D, *et al.* Somaclonal variations and their applications in horticultural crops improvement. 3 Biotech 2016; 54-71.

[24] Kumar N, Modi AR, Singh AS, *et al.* Assessment of genetic fidelity of micropropagated date palm (*Phoenix dactylifera* L.) plants by RAPD and ISSR markers assay. Physiol Mol Biol Plants 2010; 16(2): 207-13.
[http://dx.doi.org/10.1007/s12298-010-0023-9] [PMID: 23572971]

[25] Duncan RR. Tissue culture-induced variations and crop improvement. Adv Agron 1996; 58: 201-40.
[http://dx.doi.org/10.1016/S0065-2113(08)60256-4]

[26] Vazquez AM. Insight into somaclonal variations. Plant Biosyst 2001; 135: 57-62.
[http://dx.doi.org/10.1080/11263500112331350650]

[27] Doyle JJ, Doyle JL. Isolation of plant DNA from fresh tissue. Focus 1990; 12: 13-5.

[28] Varma A, Padh H, Shrivastava N. Plant genomic DNA isolation: an art or a science. Biotechnol J 2007; 2(3): 386-92.
[http://dx.doi.org/10.1002/biot.200600195] [PMID: 17285676]

[29] Sharma P, Joshi N, Sharma A. Isolation of genomic DNA from medicinal plants without liquid nitrogen. Indian J Exp Biol 2010; 48(6): 610-4.
[PMID: 20882764]

[30] Iqbal A, Ahmad I, Ahmad H, *et al.* An efficient DNA extraction protocol for medicinal plants. Int J Biosci 2013; 3: 30-5.

[31] Lade BD, Patil AS, Paikrao HM. Efficient genomic DNA extraction protocol from medicinal rich Passiflora foetida containing high level of polysaccharide and polyphenol. Springerplus 2014; 3: 457-63.

[http://dx.doi.org/10.1186/2193-1801-3-457] [PMID: 25191636]

[32]   Williams JG, Kubelik AR, Livak KJ, Rafalski JA, Tingey SV. DNA polymorphisms amplified by arbitrary primers are useful as genetic markers. Nucleic Acids Res 1990; 18(22): 6531-5.
[http://dx.doi.org/10.1093/nar/18.22.6531] [PMID: 1979162]

[33]   Bornet B, Branchard M. Nonanchored Inter Simple Sequence Repeat (ISSR) Markers: Reproducible and Specific Tool for Genome Fingerprinting. Pl Mol Biol Rep 2001; 19: 209-15.
[http://dx.doi.org/10.1007/BF02772892]

[34]   Lata H, Chandra S, Techen N, Wang YH, Khan IA. Molecular analysis of genetic fidelity in micropropagated plants of *Stevia rebaudiana* Bertoni using ISSR marker. Am J Plant Sci 2013; 4: 964-71.
[http://dx.doi.org/10.4236/ajps.2013.45119]

[35]   Nadha HK, Kumar R, Sharma RK, Anand M, Sood A. Evaluation of clonal fidelity of *in vitro* raised plants of *Guadua angustifolia* Kunth using DNA-based markers. J Med Plants Res 2011; 23: 5636-41.

[36]   Mishra T, Goyal AK, Sen S. Somatic embryogenesis and genetic fidelity study of the micropropagated medicinal spices, *Canna indica*. Horticulturae 2015; 1: 3-13.
[http://dx.doi.org/10.3390/horticulturae1010003]

[37]   Tariq A, Naz S, Shahzadi K, Ilyas S, Javed S. Study of genetic stability in *in vitro* conserved *Podophyllum hexandrum* using RAPD markers. J Anim Plant Sci 2015; 25: 1114-20.

[38]   Sharma V, Belwal N, Kamal B, Dobriyal AK, Jadon VS. Assessment of genetic fidelity of *in vitro* raised plants in *Swertia chirayita* through ISSR, RAPD analysis and peroxidase profiling during organogenesis. Braz Arch Biol Technol 2016; 59e16160389
[http://dx.doi.org/10.1590/1678-4324-2016160389]

[39]   Mahmood T, Nazar N, Abbasi BH, *et al.* Detection of somaclonal variations using RAPD fingerprinting in *Silybum marianum* (L.). J Med Plants Res 2010; 4: 1822-4.

[40]   Rawls B, Shultz KH, Dhekney S, Forrester I, Sitther V. Clonal fidelity of micropropagated *Psidium guajava* L. plants using microsatellite markers. Am J Plant Sci 2015; 6: 2385-92.
[http://dx.doi.org/10.4236/ajps.2015.614241]

Protocols used in Molecular Biology, 2020, 162-168

SUBJECT INDEX

A

Acetate 3, 4, 9, 11, 12, 62
  ammonium 62
  sodium 3, 4, 9, 12
Acidic 1, 7, 8
  nature 7, 8
  polysaccrides 1
Activity 20, 25, 26, 30, 36, 115, 117
  cysteine-aspartic acid protease 117
  endonuclease 115
  exonuclease 25, 26
  terminal transferase 30
Alkaline phosphatases 47, 48, 49, 52
Allele-specific oligonucleotide (ASO) 74
Aluminium foil 138, 139, 150
Amoebic cell(s) 129, 130, 135, 137, 138, 140, 142
  counting 137
  growing 129, 130
  live 138
  motile 140
    movements 142
  scrape 130
Amplification 16, 17, 18, 21, 22, 24, 25, 26, 30, 86, 87, 130, 156, 158
  buffer 86
  capacity 130
  cycles 16, 17, 25
  plot 18
  stings 30
Amplified 18, 74, 131, 154
  fragment length polymorphism (AFLP) 74, 154
  gene 131
  product molecules 18
Analysis 16, 107
  flow cytometric 107
  quantitative nucleic acids 16
Andor 137, 139
  EMCCD camera 137
  IQ3 acquisition software 139

Antibodies 35, 36, 39, 40, 41, 44, 45, 97, 98, 100, 104, 105, 106, 107, 108
  anti-DIG 108
  anti-phosphoprotein 39
  monoclonal 105
Antigen retrieval (AR) 44, 49, 50, 55
Antihypertensive 154
Anti-proliferative properties 120
Apoptosis 114, 115, 116, 117, 118, 120, 123
Apoptotic 115, 116, 117, 118, 120, 121, 123
  cell bodies 115
  pathway 118
  process 123
  program 123
  progression 115
  properties 115, 120, 121
  signalling 117
Arsanilic acid 52
Ascorbic acid 135
Automated cell counter for counting cells 133
Autoradiography 48
Avidin 52
  -based materials 52
  -biotin complex (ABC) 52

B

Begomoviruses 60
Binding 2, 21, 22, 24, 36, 39, 44, 45, 49, 50, 52, 70, 80, 107
  capacity 107
  efficient 22
  nonspecific primer 21
  preferential 24
Biotin forms 52
Bisacrylamide 83
Bisulfite treatment 89, 92
Bradford assay 62
Bronchopulmonarycarcinoid tumours 54
Buffers 78, 87
  enzyme dilution 87
  enzyme restriction 87

www.ingramcontent.com/pod-product-compliance
Lightning Source LLC
Chambersburg PA
CBHW041704210326
41598CB00007B/530